Heat Shock Proteins in the Immune System

Robert J. Binder • Pramod K. Srivastava
Editors

# Heat Shock Proteins in the Immune System

Springer

*Editors*
Robert J. Binder
Department of Immunology
University of Pittsburgh
School of Medicine
Pittsburgh, PA, USA

Pramod K. Srivastava
Carole and Ray Neag Comprehensive
Cancer Center and the Department
of Immunology
University of Connecticut School
of Medicine
Farmington, CT, USA

ISBN 978-3-319-69040-7 ISBN 978-3-319-69042-1 (eBook)
https://doi.org/10.1007/978-3-319-69042-1

Library of Congress Control Number: 2018934420



Printed on acid-free paper

This Springer imprint is published by the registered company Springer International Publishing AG part of Springer Nature.
The registered company address is: Gewerbestrasse 11, 6330 Cham, Switzerland

# Introduction and History

Heat shock proteins (HSPs) were discovered in 1962 [1], and, in the intervening 56 years, much of the work has focused on the role these proteins play in protecting cells from stress. Currently we know HSPs are a highly conserved class of proteins present in all species from bacteria to humans [2]. While some HSPs are constitutively expressed, others are present only upon induction. The expression of HSPs is upregulated in cells in response to various forms of stress, including deviations from physiological temperature, and deficiencies in nutrients and oxygen. To perform their role as protectors of cellular stress, many HSPs serve as chaperones, and their clients range from unfolded proteins, short peptides, and fatty acids to nucleic acids. This function is notably conserved throughout evolution [2]. In Part I, HSP-substrate interactions will be discussed. In particular, substrate binding to Hsp70 will be highlighted by Matthias Mayer. In addition, altering ligand binding through the use of molecular inhibitors and their use in treating human disease will be discussed by Len Neckers.

The properties of HSPs, including conservation through evolution, cellular ubiquity, ability to function as chaperones, and acute intracellular distribution, make them uniquely qualified to initiate and influence immune responses [3]. When HSPs gain access to the extracellular space, they are capable of initiating de novo immune responses or influencing ongoing immunity. The mechanisms by which HSPs become extracellular are the topic of Part II and will be discussed by Antonio De Maio.

The role of HSPs in the immune system was first demonstrated in a seminal paper by Srivastava and colleagues in 1986 [4]. HSPs were shown to prime antigen-specific immunity capable of rejecting tumors. Part III is dedicated to examining the role of HSPs in adaptive and innate immune responses. In the extracellular space, HSPs with their chaperoned clients provide an efficient mechanism of antigen transfer to antigen-presenting cells through a cell surface receptor. HSPs can initiate antigen cross-priming, and in addition to antigen presentation on classical MHC Ia molecules, HSPs can direct antigen presentation to nonclassical MHC Ib molecules [5]. Jacques Roberts illuminates another level of evolutionary conserved functions among HSPs by highlighting the role of HSPs in nonclassical MHC antigen presentation in the Xenopus model.

Endocytosis of HSPs by antigen-presenting cells is mediated through the receptor CD91 which has been identified as both a signaling and an endocytic receptor for gp96, Hsp70, calreticulin, and Hsp90 [6]. The role of CD91 in tumor immunosurveillance is discussed by Robert Binder. Given their ability to induce antitumor immunity, HSPs, including gp96, have been in development as immunotherapies for cancer and other diseases [7–9]. Natasa Strbo illustrates the potential uses for gp96-Ig, a novel immunotherapeutic strategy composed of a secreted form of gp96 chaperoning an antigen of choice.

In conjunction with antigen presentation, HSPs induce the maturation of antigen-presenting cells [12], and these two functions collectively result in priming of antigen-specific T cells. T cell responses are influenced by both the identity and quantity of the HSP present, with higher doses of HSPs leading to immune suppressive responses [10–12]. Songdong Meng discusses the potential for differential T cell activation by HSPs, with a particular focus on T cell responses following infection. In addition to cells of the immune system, HSPs also influence immunity by regulating noncellular components such as the extracellular matrix as discussed by Adrienne Edkins.

While extracellular HSPs are critical for developing antitumor immunity, growing tumors upregulate HSP expression as a result of intratumoral stress. These intracellular HSPs facilitate tumor cell survival and thus are tumor promoting. Hence, HSP inhibitors are being investigated for use in antitumor therapy [13]. Walter Storkus discusses the benefits and risks of using HSP inhibitors in conjunction with antitumor immunotherapies where the effects of these inhibitors may be unclear.

The mechanisms and pathways described here and discovered over the past years have led to HSPs being labeled as danger-associated molecular patterns (DAMPs) and alarmins and are implicated in phenomena such as immunogenic cell death. Collectively, our understanding of the roles of HSPs in immunity is becoming increasingly clear, and this will help to shape the development of immunotherapies for various diseases.

Abigail L. Sedlacek
Department of Immunology
University of Pittsburgh

## References

1. Ritossa F (1962) A new puffing pattern induced by temperature shock and DNP in drosophila. Experientia 18: 571–573
2. Lindquist S, Craig EA (1988) The heat-shock proteins. Annu Rev Genet 22: 631–677. doi:10.1146/annurev.ge.22.120188.003215
3. Binder RJ (2008) Heat-shock protein-based vaccines for cancer and infectious disease. Expert Rev Vaccines 7: 383–393. doi:10.1586/14760584.7.3.383
4. Srivastava PK, DeLeo AB, Old LJ (1986) Tumor rejection antigens of chemically induced sarcomas of inbred mice. Proc Natl Acad Sci USA 83: 3407–3411

5. Goyos A et al (2007) Involvement of nonclassical MHC class Ib molecules in heat shock protein-mediated anti-tumor responses. Eur J Immunol 37: 1494–1501. doi:10.1002/eji.200636570
6. Basu S, Binder RJ, Ramalingam T, Srivastava PK (2001) CD91 is a common receptor for heat shock proteins gp96, hsp90, hsp70, and calreticulin. Immunity 14: 303–313
7. Bloch O et al (2014) Heat-shock protein peptide complex-96 vaccination for recurrent glioblastoma: a phase II, single-arm trial. Neuro Oncol 16: 274–279. doi:10.1093/neuonc/not203
8. Testori A et al (2008) Phase III comparison of vitespen, an autologous tumor-derived heat shock protein gp96 peptide complex vaccine, with physician's choice of treatment for stage IV melanoma: the C-100-21 Study Group. J Clin Oncol 26: 955–962. doi:10.1200/jco.2007.11.9941
9. Wood C et al (2008) An adjuvant autologous therapeutic vaccine (HSPPC-96; vitespen) versus observation alone for patients at high risk of recurrence after nephrectomy for renal cell carcinoma: a multicentre, open-label, randomised phase III trial. Lancet 372: 145–154. doi:10.1016/s0140-6736(08)60697-2
10. Pawaria S, Binder RJ (2011) CD91-dependent programming of T-helper cell responses following heat shock protein immunization. Nat Commun 2: 521. doi:10.1038/ncomms1524
11. Chandawarkar RY, Wagh MS, and Srivastava PK (1999) The dual nature of specific immunological activity of tumor-derived gp96 preparations. J Exp Med 189: 1437–1442
12. Kinner-Bibeau LB, Sedlacek AL, Messmer MN, Watkins SC, Binder RJ (2017) HSPs drive dichotomous T-cell immune responses via DNA methylome remodelling in antigen presenting cells. Nat Commun 8:15648
13. Neckers L, Workman P (2012) Hsp90 molecular chaperone inhibitors: are we there yet? Clin Cancer Res 18: 64–76. doi:10.1158/1078-0432.ccr-11-1000

# Contents

**Part I Structure of the HSPs in Relation to Chaperoning Peptides and Proteins**

1 **Hsp70-Substrate Interactions** . . . 3
Roman Kityk and Matthias P. Mayer

2 **Molecular Chaperone Inhibitors** . . . 21
Michael A. Moses, Abbey D. Zuehlke, and Len Neckers

**Part II Exposure of HSPs to Immune Cells**

3 **Extracellular Heat Shock Proteins as Stress Communication Signals** . . . 43
Antonio De Maio

**Part III Regulation of Immune Responses by Extracellular HSPs**

4 **The Heat Shock Protein-CD91 Pathway and Tumor Immunosurveillance** . . . 63
Robert J. Binder

5 **Bridging the Gaps in the Vaccine Development: Avant-Garde Vaccine Approach with Secreted Heat Shock Protein gp96-Ig** . . . 75
Natasa Strbo

6 **Regulation of the Extracellular Matrix by Heat Shock Proteins and Molecular Chaperones** . . . 97
Natasha Marie-Eraine Boel and Adrienne Lesley Edkins

7 **Roles, Mechanisms, and Opportunities of Heat Shock Protein gp96/grp94 in Infections and Inflammation-Associated Malignancies** . . . 123
Songdong Meng and Zihai Li

**8 An Ancestral Immune Surveillance System in the Amphibian *Xenopus* Connecting Certain Heat Shock Proteins with Classical and Nonclassical MHC Class I Molecules** . . . . . . . . . . . . . . . . . . . . . . . 141
Jacques Robert, Maureen Banach, and Eva-Stina Edholm

**9 Inhibition of HSPs for Enhanced Immunity** . . . . . . . . . . . . . . . . . . . . . 157
Ronald J. Fecek, Subhara Raveendran, Manoj Chelvanambi, and Walter J. Storkus

**Index** . . . . . . . . . . . . . . . . . . . . . . . . . . . . . . . . . . . . . . . . . . . . . . . . . . . . . . 181

# Part I
# Structure of the HSPs in Relation to Chaperoning Peptides and Proteins

# Chapter 1
# Hsp70-Substrate Interactions

**Roman Kityk and Matthias P. Mayer**

**Abstract** The highly abundant and evolutionary conserved Hsp70 chaperones are central components of the cellular protein quality control system, surveilling the folding status of cellular proteins from birth at the ribosome to death through degradation. To no other chaperone families, more different functions have been assigned, and it is not surprising that Hsp70s are implicated in many developmental processes and pathological conditions. This versatility is due to the fact that Hsp70s bind tweezer-like degenerate motifs present in virtually all proteins, generally found in the hydrophobic core of the native conformation but exposed in the nascent state at the ribosome or translocation pores or upon stress-induced denaturation and aggregation. Recent years have seen much progress in understanding the molecular mechanism of this chaperone family. In this chapter, we review the current knowledge on structure, different conformational states, allostery, and regulation by co-chaperones in the context of Hsp70-substrate interaction.

**Keywords** Chaperones · Hsp70 · Hsp90 · Protein folding · Protein degradation · Protein-protein interactions · Quality control · Stress response

## Abbreviations

| | |
|---|---|
| ER | Endoplasmic reticulum |
| Hsp | Heat shock protein |
| JDP | J-domain protein, also called DnaJ proteins or Hsp40 |
| NBD | Nucleotide-binding domain |
| NEF | Nucleotide exchange factor |
| SBD | Substrate-binding domain |
| SBDα | α-Helical lid subdomain of the SBD |
| SBDβ | β-Sandwich subdomain of the SBD |
| UPR | Unfolded protein response |

R. Kityk · M. P. Mayer (✉)
Center for Molecular Biology of Heidelberg University (ZMBH),
DKFZ-ZMBH-Alliance, Heidelberg, Germany
e-mail: M.Mayer@zmbh.uni-heidelberg.de



R. J. Binder, P. K. Srivastava (eds.), *Heat Shock Proteins in the Immune System*,
https://doi.org/10.1007/978-3-319-69042-1_1

## 1.1 Introduction

The 70 kDa heat shock proteins (Hsp70s) are conserved throughout all domains of life—from bacteria to humans. As integral elements of the chaperone network in the cell, Hsp70s perform many functions, both under stress and normal conditions [1]. Hsp70s accompany proteins from "cradle to grave" as they are often among the first proteins, outside the ribosomal exit tunnel, emerging polypeptides encounter, and they are also among the last proteins before proteolytic degradation in the proteasome or lysosome [2–9] (Fig. 1.1). Nascent polypeptide chains expose hydrophobic regions which are prone to unproductive intermolecular interactions. Thus, binding of Hsp70s to newly synthesized polypeptide chains prevents aggregation and assists in de novo folding of proteins. If required, partially folded substrates can be transferred to the Hsp60 or Hsp90 systems for maturation [11–13]. Hsp70s also assist in protein translocation across membranes into endoplasmic reticulum (ER), mitochondria, and plastids [14–16]. Thereby, Hsp70s act on both sides of the membranes: cytosolic Hsp70 escorts proteins targeted for organelles in a translocation-competent state, and organellar Hsp70s (e.g. BiP in the ER, mtHsp70 in mitochondria) bind the emerging substrates at the translocation pore and promote the transport into the lumen of the organelle [16–19]. Stress conditions and macromolecular crowding in the cell promote protein aggregation, which can be prevented or counteracted by disaggregation activity of Hsp70s in cooperation with other chaperone families [20–22].

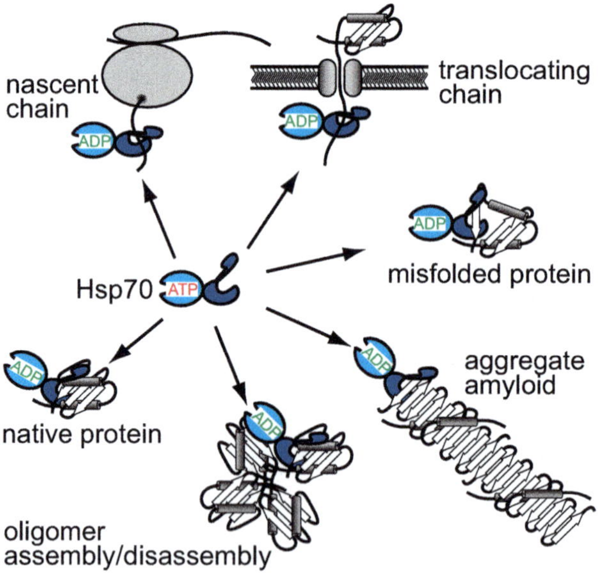


**Fig. 1.1** Multiple functions of Hsp70 chaperones in the cell. The cartoon illustrates the different types of substrates and protein conformations encountered by Hsp70s, including extended conformations in de novo folding and translocation; partially folded, molten globule-like and misfolded conformations; native proteins; protein assemblies; and amorphous and amyloidic aggregates (figure from [10])

Given the wide range of tasks performed by Hsp70s in the cell, it is not surprising that a number of diseases are linked to the activity of this chaperone family, including cancer and neurodegenerative disorders to give just two examples. It was reported that Hsp70 levels in cancer cells are elevated, increasing their viability and drug resistance [23–28]. Hsp70 controls the activity of wild-type and mutant tumour suppressor p53, thereby counteracting the induction of apoptosis [29–31]. On the other side, overexpression of Hsp70 can overcome negative symptoms of neurodegenerative diseases [32–34]. Such an involvement of Hsp70 in oncogenesis and neurodegeneration processes highlights the importance of understanding the molecular mechanisms which govern the functioning of Hsp70.

## 1.2 Hsp70 Functional Cycle

Hsp70s consist of a 45 kDa N-terminal nucleotide-binding domain (NBD), which possesses low intrinsic ATPase activity, and a 25 kDa C-terminal substrate-binding domain (SBD), connected via a conserved hydrophobic linker. At the heart of Hsp70s' chaperone activity is the ATPase cycle, in which they oscillate between two distinct functional states (Fig. 1.2). The ATP state is characterized by low affinity and high exchange rates for polypeptide substrates, while in the ADP-bound state, Hsp70s have high affinity and low on- and off-rates for substrates [35, 36]. During their chaperone cycle, Hsp70s are aided by co-chaperones—J-domain proteins (JDPs), some of which also interact with substrates—and nucleotide exchange factors (NEFs).

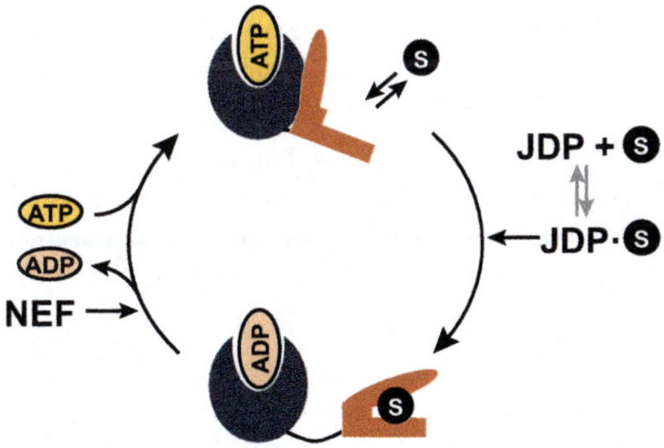


**Fig. 1.2** Functional cycle of Hsp70 proteins. Hsp70s cycle between two states—the ATP-bound state with low affinity for substrates (S) and the ADP-bound state with high affinity for substrates. The cycle is controlled by internal allostery of the protein and modulated by co-chaperones. J-domain proteins (JDPs) facilitate the transfer of the substrate onto Hsp70 and couple substrate binding with ATP hydrolysis, resulting in efficient trapping of the substrate. Nucleotide exchange factors (NEFs) catalyse ADP release, accelerating rebinding of ATP and leading to a conformational change of the chaperone and release of the substrate

The chaperone cycle starts with a rapid and transient association of Hsp70·ATP with the substrate. JDPs catalyse this step by stimulating the low intrinsic ATPase activity of Hsp70s in synergism with the substrate some 100–1000-fold, resulting in trapping of the substrate [37–39]. The release of the substrate requires the exchange of ADP for ATP, as this becomes the rate-limiting step of the ATPase cycle in the presence of JDPs. NEFs accelerate ADP release [40–43], and subsequent binding of ATP displaces the NEF and converts Hsp70 into the low affinity state, leading to substrate release and the reset of the cycle.

## 1.3 Structural Basis for Hsp70-Substrate Interactions

The Hsp70s' capability to perform diverse tasks relies on their ability to bind in an ATP-dependent manner short degenerative peptide motifs enriched in hydrophobic and positively charged residues, which can be found on average every 30–40 residues within virtually all proteins, except for intrinsically disordered proteins [44]. Most of the structural information about the SBD-substrate interactions is based on the X-ray structure of the SBD from *E. coli* Hsp70 homolog DnaK co-crystalized with a model substrate peptide [45]. The substrate-binding domain consists of a β-sandwich subdomain (SBDβ), harbouring two β-sheets with four strands each, and an α-helical subdomain (SBDα) containing five helices—A, B, C, D, and E (Fig. 1.3b). The substrate-binding pocket is formed by the two twisted β-sheets and two sets of loops—$L_{1,2}$ and $L_{3,4}$—which form a cradle for the substrate backbone, stabilized by a second layer of loops, $L_{4,5}$ and $L_{5,6}$. Helices A and B pack against and stabilize $L_{4,5}$ and $L_{1,2}$, and the distal part of helix B forms a salt bridge and two hydrogen bonds with $L_{3,4}$ and $L_{5,6}$ and thus acts like a lid and a latch, which closes over a bound peptide substrate [46]. Whether helices C, D and, E have an additional function other than stabilizing the distal part of helix B is unclear. The last 31 residues at the C terminus were cleaved off prior to crystallization due to the high degree of flexibility in this region. In *E. coli*, the C terminus of the SBD was suggested to be involved in interactions with substrates [47]. In eukaryotic organisms, the C terminus of cytosolic Hsp70 homologs ends in the EEVD motif, enabling interactions of Hsp70s with TPR domain containing co-chaperones [48].

The crystal structure reveals that peptides bind to the SBD in an extended conformation. Five amino acids are engaged in two main types of interaction with DnaK. First, hydrogen bonds are formed between the substrate and loops $L_{1,2}$ and $L_{3,4}$, in particular involving the backbone of the substrate, explaining the preference for natural peptides made of L-amino acids over peptides made of D-amino acids [49]. The second type of interactions between peptide and DnaK are van de Waals interactions, between hydrophobic side chains of the substrate and hydrophobic residues lining the substrate-binding cavity. Additionally, the surface surrounding of the substrate-binding cleft is negatively charged, which explains why DnaK prefers substrates containing a central core of hydrophobic amino acids, flanked by positively charged residues [44].

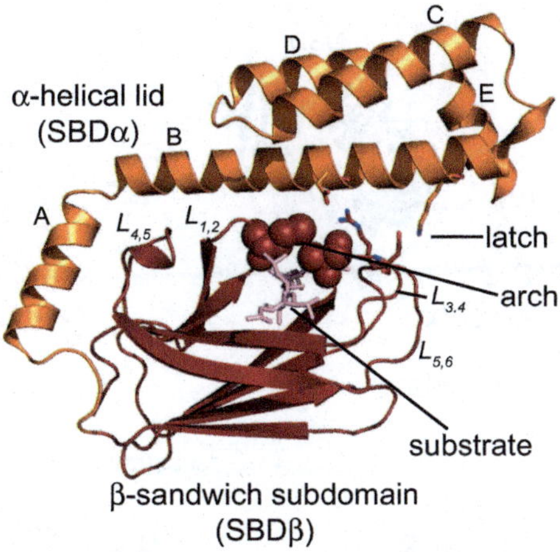


**Fig. 1.3** Secondary structure representation of the SBD from *E. coli* DnaK in the substrate-bound state (PDB code, 1DKX). The SBD consists of a β-sandwich subdomain and an α-helical lid which closes over it. The lid in the closed state is stabilized by the electrostatic latch formed between helix B and loops $L_{3,4}$ and $L_{5,6}$. Peptide substrate is bound in an extended conformation and is enclosed in the substrate-binding cleft by the hydrophobic arch formed by the inner loops $L_{1,2}$ and $L_{3,4}$ and stabilized by the outer loops $L_{4,5}$ and $L_{5,6}$

Efficient substrate release from the Hsp70-SBD is achieved via ATP binding by the Hsp70-NBD. Crystal structures of a two-domain construct of DnaK·ATP provided the molecular basis for the low substrate affinity and for interdomain communication within Hsp70s [50, 51]. ATP binding induces a rotation of NBD subdomains as compared to the ADP state and in the SBD SBDβ and α-helical lid detach from each other and dock to different parts of the NBD. Thus, the lid does not cover the substrate-binding channel anymore, and the SBDβ is stabilized in a wide open conformation with $L_{1,2}$ and $L_{4,5}$ shifted towards the NBD, consistent with low affinity and high dissociation rates for the substrate in the ATP state (Fig. 1.4). In addition, the hydrophobic pocket in the SBDβ, which binds the central hydrophobic residue of the substrate in the structure of the SBD-substrate peptide complex, is diminished in width in the ATP-bound state. All of these ATP-induced structural rearrangements result in efficient substrate release. The interface between NBD and SBDβ contains an extensive H-bond network, which was shown to be the heart of the allosteric mechanism in Hsp70s [52, 53].

## 1.4 Mechanism of Action of Hsp70 Chaperones

Different models have been proposed to explain the effects of Hsp70s on substrate proteins. In the "kinetic partitioning" model, the chaperone does not affect the conformation of unfolded or misfolded substrates but only binds transiently to their exposed hydrophobic polypeptide stretches, preventing intermolecular

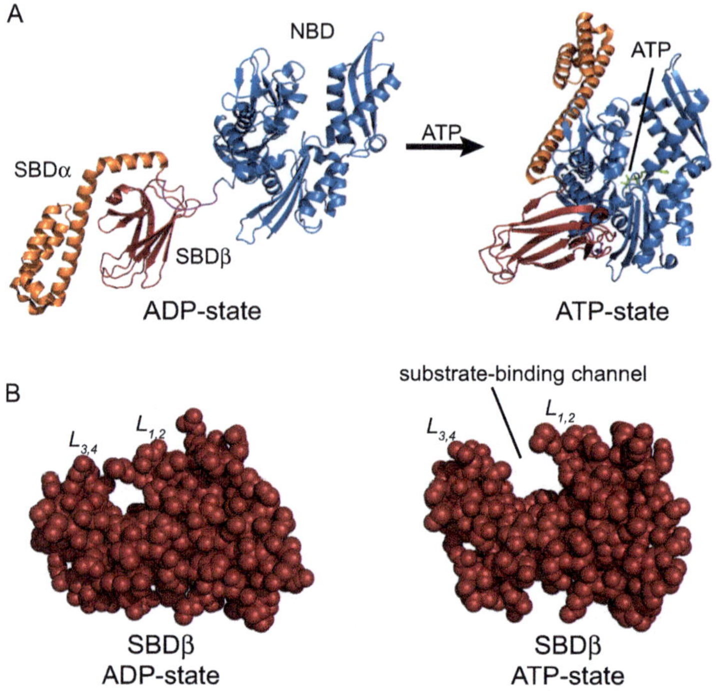


**Fig. 1.4** Structural basis for the ATP-induced substrate release from the Hsp70s. (**a**) Binding of ATP leads to dramatic structural changes in DnaK. On the left—NMR structure of DnaK·ADP (PDB code, 2KHO), which demonstrates that in the ADP-bound state, NBD and SBD are in disjoined conformation; on the right, structure of DnaK·ATP (PDB code, 4B9Q), in which the SBD (both SBDα and SBDβ) is docked onto NBD and stabilized in the open conformation. (**b**) Filled space representations of the substrate-bound SBDβ (PDB code, 1DKX; represents the ADP state) and SBDβ in the DnaK·ATP structure (PDB code, 4B9Q); in the ATP state, loops $L_{1,2}$ and $L_{3,4}$ do not enclose the substrate-binding pocket leaving the substrate-binding channel widely open, which results in low affinity and high exchange rates for the substrate

homotypic association of these hydrophobic sites, thereby decreasing the pool of the free, aggregation-prone species in solution. In contrast to folding, protein aggregation is a concentration-dependent process, and thus, "kinetic partitioning" would facilitate the folding pathway by decreasing the rates of aggregation. Such a mechanism explains how Hsp70s may prevent protein aggregation and promote de novo folding of nascent polypeptides, but does not seem to be operative for refolding of proteins which are trapped in a non-native conformation. An alternative model suggests that Hsp70s can induce unfolding of misfolded substrates through binding-release cycles, allowing them to refold subsequently. Evidence was provided that the DnaK system unfolds a misfolded model substrate, a variant of firefly luciferase, prior to refolding [54]. Evidence was also provided that DnaK induces local unfolding in a native protein [55]. A different study suggested that Hsp70s can provide the surface on which the bound substrates can sample different conformations [56].

For protein translocation across membranes, a model called "entropic pulling" was proposed [57, 58]. Briefly, as the polypeptide chain emerges from the translocon, it interacts with Hsp70 targeted by membrane-bound JDPs. Due to the excluded volume effects (because of the close proximity to membrane), the number of possible conformations of the polypeptide bound to Hsp70 is rather limited. However, as the number of translocated residues increases, the distance between Hsp70 and the membrane becomes larger, thereby increasing the conformational freedom of the translocating peptide and the total entropy of the system. This increase in entropy generates a force which drives the translocation across the membrane. Since Hsp70-binding motifs are found in proteins on average every 30–40 amino acids, the "entropic pulling" mechanism would ensure that long polypeptide chains are efficiently translocated across the membrane. A similar mechanism could be also applicable for the disaggregation function of Hsp70s in particular for the extraction of polypeptide chains from aggregates.

## 1.5 Interactions of Hsp70 Chaperones with Protein Substrates

In the cell, Hsp70s encounter mainly protein substrates. Model peptide substrates are useful tools for investigation of the molecular mechanism of Hsp70-substrate interactions; however, there are certain differences in Hsp70s' interaction with peptide and protein substrates. First, peptide substrates stimulate the ATPase activity of Hsp70s much less efficiently than protein substrates, and synergism with JDPs is not observed [38]. Second, biochemical data suggests that the SBD of Hsp70s adopts different conformations when interacting with peptide and protein substrates. The α-helical lid is closed completely when Hsp70s interact with peptide substrates, while it does not close completely over protein substrates [10, 59]. Thus, Hsp70s do not necessarily interact only with extended conformations of protein substrates, in contrast to what was suggested from the crystal structure of the SBD with a substrate peptide. This mode of interaction may be crucial for protein disaggregation and interaction with native proteins.

During stress conditions (e.g. heat stress), proteins can misfold and aggregate, which can have detrimental consequences for the cell. In prokaryotes, organelles of prokaryotic origin, as well as yeast and plants, Hsp70s can disassemble protein aggregates in collaboration with Hsp100s, which are toroidal hexameric AAA+ proteins. In this case, Hsp70 acts both upstream and downstream of Hsp100, helping to extract single polypeptide chains from the aggregate, and then promotes their refolding after being unfolded by Hsp100 [60–62]. In metazoans, which lack Hsp100s, protein disaggregation is performed by Hsp70 in cooperation with Hsp110 chaperones, which are relatives of Hsp70s and act as NEFs for Hsp70s [63–65]. Recently, the human Hsp70-Hsp40-Hsp110 system was demonstrated to dissolve Parkinson's-related α-synuclein amyloid fibres [66]. In this case, the Hsp70 machinery is able to fragment the fibres and thereby generates new free ends, which seem to be the preferential sites for Hsp70 action in α-synuclein fibrils, resulting in acceleration of fibril depolymerization.

Hsp70s can interact not only with unfolded or misfolded polypeptides but also with natively folded or near-native proteins, performing regulatory functions in the cell. The best studied example is the interaction of DnaK with the heat shock transcription factor $\sigma^{32}$ in *E. coli* [55, 67]. Based on the homology models of $\sigma^{32}$ to other σ factors, it was suggested that $\sigma^{32}$ exists in a compact and an extended conformation which are in equilibrium with each other, with the latter preferentially interacting with DnaK. It was suggested that the conformational change of $\sigma^{32}$ is the rate-limiting step for DnaK binding and that binding of DnaK shifts the conformational equilibrium of $\sigma^{32}$ towards the extended conformation. DnaK binding results in unfolding in a defined region of $\sigma^{32}$, which was proposed to facilitate the degradation of $\sigma^{32}$ by the protease FtsH [55]. Upon heat shock, DnaK gets titrated away from $\sigma^{32}$ by other misfolded proteins, resulting in stabilization of $\sigma^{32}$ and expression of the heat shock genes, thereby providing the foundations for the regulation of the heat shock response in *E. coli.* In eukaryotes, Hsp70 is also involved in the regulation of heat shock response in the nuclear-cytoplasmic compartment, since it was found to interact with heat shock factor 1 (HSF1) during the attenuation phase of the heat shock response [68, 69]. Similarly, the ER Hsp70 homolog BiP was proposed to be the regulator of the unfolded protein response (UPR). Initiation of the UPR through two out of three known pathways involves homodimerization of ER membrane-embedded receptors—IRE1 and PERK. Although the UPR activation mechanism is still under debate, it was suggested that BiP can bind to IRE1 and PERK and prevent their homodimerization, thereby supressing the UPR [70]. Under stress conditions, the increasing amount of the unfolded proteins in the ER outcompetes IRE1 and PERK for BiP binding, allowing the UPR receptors to dimerize. Dimerization of the IRE1 results in the activation of its RNase activity, required for the noncanonical splicing of the mRNA coding for XBP1, a transcription factor which drives the expression of the genes encoding the proteins (e.g. chaperones and BiP itself) that counteract the ER stress. Dimerized PERK phosphorylates eIF2α, globally decreasing translation levels and protein influx into the ER.

Additional examples of native proteins interacting with the *E. coli* Hsp70 system are DNA replication initiator proteins like λP [71], RepA [72], and RepE [73]. Thereby, the Hsp70 system disassembles homodimers (RepA and RepE) or a heteromeric complex (λP-DnaB) to activate DNA replication. There are many more examples of Hsp70s interacting with native or near-native proteins in eukaryotic cells in addition to the above-mentioned examples. Thereby, Hsp70s often cooperate with Hsp90 chaperones. These so-called clients of the Hsp70-Hsp90 chaperone machinery include many transcription factors, e.g. steroid hormone receptors and p53, many kinases, and many other proteins important for a large variety of cellular functions [74, 75]. For some of these clients, the interaction with Hsp70 and Hsp90 seem to be restricted to folding and maturation, but others appear to require chaperone assistance throughout their entire lifespan. Hsp70 and Hsp90 are thereby involved in coupling environmental conditions to cellular and developmental signals controlling cell homeostasis, proliferation, differentiation, and cell death.

In addition to interactions with folded proteins, Hsp70s can also disassemble protein complexes as mentioned above. A classic example for this aspect of Hsp70 function is the disassembly of clathrin from clathrin-coated endocytic vesicles by the constitutive cytosolic Hsc70 [76]. The clathrin triskelia undergo conformational fluctuations exposing Hsc70-interacting motifs. Hsc70 binding stabilizes triskelia in a conformation that induces a strain in the clathrin baskets. Binding of multiple Hsc70 molecules to triskelia increases the conformational strain to a critical level ultimately leading to the cooperative disassembly of the clathrin coat [77, 78].

Hsp70s play also an important role at different stages of viral infections, from cell membrane penetration to capsid assembly [79]. Although it is not completely understood how Hsp70s reach the cell exterior, there is evidence that some viruses require surface-exposed Hsp70s in order to enter the host cell, including rotavirus, coxsackievirus A9 (CAV-9), dengue virus, and human T cell lymphotropic virus type 1 (HTLV-1) [80–84]. Additionally, Hsp70s were shown to be involved in the disassembly of the viral coats of polyomaviruses, papillomaviruses, and reoviruses [85, 86]. Hsp70s were also demonstrated to regulate the DNA replication process of viruses. As mentioned above, the *E. coli* DnaK system releases the helicase DnaB from the complex with λP, which is required for the initiation of DNA replication of bacteriophage λ [87, 88]. There are also reports that Hsp70 system remodels replication pre-initiation complexes of eukaryotic viruses [89–92]. Lastly, the Hsp70 system plays an important role in virion assembly of some viruses, not only during folding of the capsid proteins [93–95].

## 1.6 Role of Co-Chaperones

Hsp70s usually do not act on their own but are supported by a number of co-chaperones. In the simplest case, the set of required co-chaperones includes a JDP and a NEF, which stimulate ATP hydrolysis and ADP-to-ATP exchange, respectively.

JDPs are proposed to be targeting factors for Hsp70s, which bind substrates themselves and transfer them to Hsp70 or are located in close proximity where Hsp70 substrates emerge like at the ribosome or translocon. Cells express a wide spectrum of JDPs, which have diverse architecture and functions and are divided into three classes, A, B, and C, according to their domain composition [96]. All of them possess a J-domain with the conserved HPD motif, critical for the interaction with Hsp70s. Interestingly, the number of different JDPs in cells is generally much higher than the number of Hsp70s. For example, humans in total possess 11 Hsp70s but 47 JDPs, or rather 53, if all splice variants are counted separately. Taking into account the diversity of JDPs, it is not surprising that besides their basic function of substrate delivery and coupling it with ATP hydrolysis, they play an important role in determining and shaping many Hsp70 functions in the cell. As mentioned above,

JDPs can also target Hsp70 to locations where its activity is required. In eukaryotes, JDPs (e.g. zuotin in yeast, MPP11 in human cells) are one of the components of the ribosome-associated complex (RAC), targeting cytosolic Hsp70s to the ribosome for early steps of de novo folding during protein synthesis [4, 97–99]. The J-domain-containing Pam16-Pam18 complex targets the mitochondrial Hsp70 to the translocon in the inner membrane of mitochondria thereby supporting protein import into the mitochondrial matrix [19, 100]. Similarly, Sec63, which is a transmembrane JDP in the endoplasmic reticulum (ER) membrane with the J-domain in the ER lumen, targets BiP to the translocon and facilitates protein transport into the ER. It is also known that JDPs are involved in ER-associated degradation (ERAD; e.g. ERdj4 and ERdj5) [101–103] or can possess ubiquitin-interacting domains, providing a link between the Hsp70 system and the protein degradation machinery in the cell [104]. Recently, JDPs of different classes were shown to synergistically activate the disaggregation function of the metazoan Hsp70 system [64]. This study revealed that JDPs from classes A and B recognize protein aggregates of different sizes, allowing efficient Hsp70 targeting to aggregates on different stages of the disaggregation process to compensate for the heterogeneity of aggregates. The JDPs from classes A and B also form transient complexes in vitro and in vivo, and this complex formation is necessary for synergistic disaggregation. If such complexes are more common, the already large number of JDPs will be potentiated through combinatorial complex formation, most likely enlarging the substrate spectrum of Hsp70s.

There are several structurally unrelated groups of NEFs. The first NEF to be discovered was GrpE from *E. coli*. In eukaryotes, however, GrpE homologs can only be found in mitochondria and chloroplasts. In other compartments, it is replaced by NEFs from the structurally unrelated BAG, HspBP1, and Hsp110/Hsp170 families [105–108]. Hsp110s and BAG proteins are particularly interesting, because they have additional roles apart from accelerating nucleotide exchange, which shape the functional landscape of the Hsp70 system in the cell.

BAG (Bcl-2-associated athanogene) proteins are very diverse, modular, cytosolic proteins that are characterized by the presence of one or several so-called BAG domains, a three-helix bundle, which is essential for interaction with Hsp70s and provides the NEF function [109]. Some Bag proteins have been proposed to divert Hsp70 substrates to degradation. Bag1 contains a ubiquitin-like domain and is, in addition, ubiquitinated by the E3 ligase and Hsp70 co-chaperone CHIP, both of which promote binding to the proteasome, suggesting a role in linking the Hsp70 system to the proteasome degradation system [110]. Bag3 facilitates the interaction of ubiquitinated and non-ubiquitinated Hsp70 substrates with the p62 (or NRB1) adaptor protein on the phagophore membrane, providing a connection of the Hsp70 system towards autophagy [111–113]. Bag6/Bat3/Scythe interacts with pro-apoptotic factors, enhancing polyubiquitination and promoting their proteasomal degradation [114–117]. BAG proteins are also involved in other functions, including prevention of protein aggregation and protein folding [109, 117].

Hsp110/Hsp170 proteins, which act as Hsp70 NEFs in the nuclear-cytoplasmic compartment (Hsp110) and in the ER (Hsp170), are Hsp70 homologs themselves, also consisting of a nucleotide-binding domain, a β-sandwich subdomain, and an α-helical domain, which were crystalized in a conformation highly similar to the ATP-bound open conformation of Hsp70s [118–122]. Hsp110s also were shown to interact with peptide and protein substrates [66, 123, 124]. Whether they are also able to undergo similar conformational changes as Hsp70s is not clear. Hsp110 proteins of higher eukaryotes are involved in protein disaggregation and α-synuclein fibril fragmentation together with Hsp70 [63–66].

There is also an anti-NEF called Hip (Hsp70-interacting protein) which binds to the NBD of Hsp70 at a location that overlaps with the binding sites for NEFs but does not accelerate nucleotide exchange [125]. Since nucleotide exchange is rate-limiting for substrate release, NEFs and anti-NEF regulate the lifetime of the Hsp70-substrate complex, which is likely to be important for efficient folding or transfer to other chaperone systems or the degradation system.

As mentioned above, the C terminus of eukaryotic cytosolic Hsp70s possesses an EEVD motif which is the interaction site for TPR domain containing co-chaperones. One of such co-chaperones is Hop (Hsp70–Hsp90-organizing protein; Sti1 in yeast). Hop contains three TPR domains and can simultaneously interact with the GTIEEVD motif of Hsp70 and the MEEVD motif of Hsp90, resulting in a ternary complex. Thus, it is suggested that Hop couples the functional cycles of Hsp70 and Hsp90 and facilitates loading of certain partially folded intermediates from Hsp70 onto the Hsp90 machinery [126, 127]. Such a coupling of both cycles was demonstrated to be critical for the maturation of steroid hormone receptors. Folding of the ligand-binding domain of glucocorticoid receptor was shown to require a handover from the Hsp70 onto the Hsp90 system and was dependent on the presence of Hop [128, 129]. Based on the study of the interaction between Hsp90 and Alzheimer-linked protein Tau, a mechanism of the Hsp70-Hsp90 cooperation was proposed [130]. Hsp70 was suggested to bind early to nascent polypeptide chains which still expose larger hydrophobic stretches constituting high-affinity Hsp70-binding motifs. As polypeptide folding progresses, Hsp70-binding sites become buried and inaccessible for further interaction with Hsp70, leaving, however, scattered hydrophobic residues exposed. The latter are recognized by the extended substrate-binding surface of Hsp90, which facilitates the late stages of the folding process, resulting in the maturation of the substrate protein.

Another co-chaperone that interacts with Hsp70 via the C-terminal EEVD motif is CHIP (C terminus of Hsc70-interacting protein), which possesses E3 ligase activity [131–133]. According to the current view, CHIP stochastically promotes ubiquitination of substrates that are in complex with Hsp70, preferentially targeting such misfolded proteins to the proteasome which spend longer time bound to the chaperone (i.e. undergo more binding-release cycles) and presumably are difficult to fold or stuck in a non-refoldable misfolded state [134]. Thereby, substrates which can be folded/refolded within a short time frame by Hsp70 have decreased likelihood of being targeted for degradation. Taken together, the interplay between Hsp70 and the plethora of its co-chaperones is at the core of the triage decision determining the fate of its substrates: refolding to the native state, transferring to Hsp90, and targeting for degradation by the proteasome or through autophagy.

## 1.7 Concluding Remarks

Hsp70s interact with a large number of substrate proteins in vivo. For *E. coli* DnaK, more than 700 in vivo substrates were identified [135]. Some of these proteins are transferred onto the Hsp60 machinery for further folding steps, while the conformational maintenance of others depends on the continuous interaction with DnaK. Thus, DnaK was proposed to be a central hub of the chaperone network in the *E. coli* cell. Proteins in eukaryotic cells are on average larger as compared to prokaryotic organisms and hence demand much more attention from the protein quality control network. Thus, in the course of evolution, the Hsp70 system developed into a very sophisticated machinery, which determines the fate and regulates the activity of many proteins.

Considering the central role which Hsp70s play in the proteostasis network in the cell, it is not surprising that deregulation of the Hsp70 activity leads to pathophysiological processes, particularly cancer and neurodegeneration. Therefore, understanding the basic molecular principles how the Hsp70 machinery functions is important from a medical perspective, and the Hsp70-substrate interactions are one of the key aspects here. Although the basic principles of the Hsp70-peptide interactions are rather well understood, the knowledge on the mechanism of Hsp70-protein interactions begins only to emerge. One of the limitations in this aspect is the lack of a high-resolution structure of Hsp70-protein substrate complexes. Another open question is how the flexibility of the SBD affects substrate specificity of different Hsp70s, since different Hsp70 homologs seem to have different conformational plasticities within their substrate-binding cleft, resulting in different kinetic parameters of Hsp70-substrate interactions [136]. Hsp70s in eukaryotic organisms are targets for multiple post-translational modifications, which might provide yet another complexity level of the regulation of Hsp70-substrate interactions and should be addressed in the future.

## References


1. Richter K, Haslbeck M, Buchner J (2010) The heat shock response: life on the verge of death. Mol Cell 40:253–266
2. Deuerling E, Schulze-Specking A, Tomoyasu T, Mogk A, Bukau B (1999) Trigger factor and DnaK cooperate in folding of newly synthesized proteins. Nature 400:693–696
3. Teter SA, Houry WA, Ang D, Tradler T, Rockabrand D, Fischer G, Blum P, Georgopoulos C, Hartl FU (1999) Polypeptide flux through bacterial Hsp70. Cell 97:755–765
4. Preissler S, Deuerling E (2012) Ribosome-associated chaperones as key players in proteostasis. Trends Biochem Sci 37:274–283
5. Shiber A, Ravid T (2014) Chaperoning proteins for destruction: diverse roles of Hsp70 chaperones and their co-chaperones in targeting misfolded proteins to the proteasome. Biomol Ther 4:704–724
6. Meimaridou E, Gooljar SB, Chapple JP (2009) From hatching to dispatching: the multiple cellular roles of the Hsp70 molecular chaperone machinery. J Mol Endocrinol 42:1–9
7. Xilouri M, Stefanis L (2015) Chaperone mediated autophagy to the rescue: a new-fangled target for the treatment of neurodegenerative diseases. Mol Cell Neurosci 66:29–36

8. Cuervo AM, Wong E (2014) Chaperone-mediated autophagy: roles in disease and aging. Cell Res 24:92–104
9. Ulbricht A, Hohfeld J (2013) Tension-induced autophagy: may the chaperone be with you. Autophagy 9:920–922
10. Schlecht R, Erbse AH, Bukau B, Mayer MP (2011) Mechanics of Hsp70 chaperones enables differential interaction with client proteins. Nat Struct Mol Biol 18:345–351
11. Langer T, Pfeifer G, Martin J, Baumeister W, Hartl FU (1992) Chaperonin-mediated protein folding: GroES binds to one end of the GroEL cylinder, which accommodates the protein substrate within its central cavity. EMBO J 11:4757–4765
12. Young JC, Agashe VR, Siegers K, Hartl FU (2004) Pathways of chaperone-mediated protein folding in the cytosol. Nat Rev Mol Cell Biol 5:781–791
13. Li J, Soroka J, Buchner J (2012) The Hsp90 chaperone machinery: conformational dynamics and regulation by co-chaperones. Biochim Biophys Acta 1823:624–635
14. Matlack KE, Plath K, Misselwitz B, Rapoport TA (1997) Protein transport by purified yeast Sec complex and Kar2p without membranes. Science 277:938–941
15. Wiedemann N, Frazier AE, Pfanner N (2004) The protein import machinery of mitochondria. J Biol Chem 279:14473–14476
16. Neupert W, Brunner M (2002) The protein import motor of mitochondria. Nat Rev Mol Cell Biol 3:555–565
17. Young JC, Hoogenraad NJ, Hartl FU (2003) Molecular chaperones Hsp90 and Hsp70 deliver preproteins to the mitochondrial import receptor Tom70. Cell 112:41–50
18. Flores-Pérez Ú, Jarvis P (2013) Molecular chaperone involvement in chloroplast protein import. Biochim Biophys Acta 1833:332–340
19. Chacinska A, Koehler CM, Milenkovic D, Lithgow T, Pfanner N (2009) Importing mitochondrial proteins: machineries and mechanisms. Cell 138:628–644
20. Mogk A, Kummer E, Bukau B (2015) Cooperation of Hsp70 and Hsp100 chaperone machines in protein disaggregation. Front Mol Biosci 2:22
21. Sousa R (2014) Structural mechanisms of chaperone mediated protein disaggregation. Front Mol Biosci 1:12
22. Mogk A, Tomoyasu T, Goloubinoff P, Rüdiger S, Röder D, Langen H, Bukau B (1999) Identification of thermolabile Escherichia coli proteins: prevention and reversion of aggregation by DnaK and ClpB. EMBO J 18:6934–6949
23. Nylandsted J, Rohde M, Brand K, Bastholm L, Elling F, Jäättelä M (2000) Selective depletion of heat shock protein 70 (Hsp70) activates a tumor-specific death program that is independent of caspases and bypasses Bcl-2. Proc Natl Acad Sci U S A 97:7871–7876
24. Nylandsted J, Gyrd-Hansen M, Danielewicz A, Fehrenbacher N, Lademann U, Høyer-Hansen M, Weber E, Multhoff G, Rohde M, Jäättelä M (2004) Heat shock protein 70 promotes cell survival by inhibiting lysosomal membrane permeabilization. J Exp Med 200:425–435
25. Ding L, He S, Sun X (2014) HSP70 desensitizes osteosarcoma cells to baicalein and protects cells from undergoing apoptosis. Apoptosis 19:1269–1280
26. Guo F, Sigua C, Bali P, George P, Fiskus W, Scuto A, Annavarapu S, Mouttaki A, Sondarva G, Wei S, Wu J, Djeu J, Bhalla K (2005) Mechanistic role of heat shock protein 70 in Bcr-Abl-mediated resistance to apoptosis in human acute leukemia cells. Blood 105:1246–1255
27. Lianos GD, Alexiou GA, Mangano A, Mangano A, Rausei S, Boni L, Dionigi G, Roukos DH (2015) The role of heat shock proteins in cancer. Cancer Lett 360:114–118
28. Zorzi E, Bonvini P (2011) Inducible hsp70 in the regulation of cancer cell survival: analysis of chaperone induction, expression and activity. Cancers (Basel) 3:3921–3956
29. Fourie AM, Hupp TR, Lane DP, Sang BC, Barbosa MS, Sambrook JF, Gething MJ (1997) HSP70 binding sites in the tumor suppressor protein p53. J Biol Chem 272:19471–19479
30. King FW, Wawrzynow A, Hohfeld J, Zylicz M (2001) Co-chaperones Bag-1, Hop and Hsp40 regulate Hsc70 and Hsp90 interactions with wild-type or mutant p53. EMBO J 20:6297–6305
31. Zylicz M, King FW, Wawrzynow A (2001) Hsp70 interactions with the p53 tumour suppressor protein. EMBO J 20:4634–4638

32. Bonini NM (2002) Chaperoning brain degeneration. Proc Natl Acad Sci U S A 99(Suppl 4):16407–16411
33. Auluck PK, Chan HYE, Trojanowski JQ, Lee VMY, Bonini NM (2002) Chaperone suppression of alpha-synuclein toxicity in a Drosophila model for Parkinson's disease. Science 295:865–868
34. Klucken J, Shin Y, Masliah E, Hyman BT, McLean PJ (2004) Hsp70 reduces alpha-synuclein aggregation and toxicity. J Biol Chem 279:25497–25502
35. Mayer MP, Schröder H, Rüdiger S, Paal K, Laufen T, Bukau B (2000) Multistep mechanism of substrate binding determines chaperone activity of Hsp70. Nat Struct Biol 7:586–593
36. Schmid D, Baici A, Gehring H, Christen P (1994) Kinetics of molecular chaperone action. Science 263:971–973
37. Karzai AW, McMacken R (1996) A bipartite signaling mechanism involved in DnaJ-mediated activation of the Escherichia coli DnaK protein. J Biol Chem 271:11236–11246
38. Laufen T, Mayer MP, Beisel C, Klostermeier D, Mogk A, Reinstein J, Bukau B (1999) Mechanism of regulation of hsp70 chaperones by DnaJ cochaperones. Proc Natl Acad Sci U S A 96:5452–5457
39. Barouch W, Prasad K, Greene L, Eisenberg E (1997) Auxilin-induced interaction of the molecular chaperone Hsc70 with clathrin baskets. Biochemistry 36:4303–4308
40. Liberek K, Marszalek J, Ang D, Georgopoulos C (1991) Escherichia coli DnaJ and GrpE heat shock proteins jointly stimulate ATPase activity of DnaK. Proc Natl Acad Sci U S A 88(7):2874–2878
41. Packschies L, Theyssen H, Buchberger A, Bukau B, Goody RS, Reinstein J (1997) GrpE accelerates nucleotide exchange of the molecular chaperone DnaK with an associative displacement mechanism. Biochemistry 36:3417–3422
42. Gässler CS, Wiederkehr T, Brehmer D, Bukau B, Mayer MP (2001) Bag-1M accelerates nucleotide release for human Hsc70 and Hsp70 and can act concentration-dependent as positive and negative cofactor. J Biol Chem 276:32538–32544
43. Brehmer D, Rüdiger S, Gässler CS, Klostermeier D, Packschies L, Reinstein J, Mayer MP, Bukau B (2001) Tuning of chaperone activity of Hsp70 proteins by modulation of nucleotide exchange. Nat Struct Biol 8:427–432
44. Rüdiger S, Germeroth L, Schneider-Mergener J, Bukau B (1997) Substrate specificity of the DnaK chaperone determined by screening cellulose-bound peptide libraries. EMBO J 16:1501–1507
45. Zhu X, Zhao X, Burkholder WF, Gragerov A, Ogata CM, Gottesman ME, Hendrickson WA (1996) Structural analysis of substrate binding by the molecular chaperone DnaK. Science 272:1606–1614
46. Mayer MP, Rüdiger S, Bukau B (2000) Molecular basis for interactions of the DnaK chaperone with substrates. Biol Chem 381:877–885
47. Smock RG, Blackburn ME, Gierasch LM (2011) Conserved, disordered C terminus of DnaK enhances cellular survival upon stress and DnaK in vitro chaperone activity. J Biol Chem 286:31821–31829
48. Demand J, Lüders J, Höhfeld J (1998) The carboxy-terminal domain of Hsc70 provides binding sites for a distinct set of chaperone cofactors. Mol Cell Biol 18:2023–2028
49. Rüdiger S, Schneider-Mergener J, Bukau B (2001) Its substrate specificity characterizes the DnaJ co-chaperone as a scanning factor for the DnaK chaperone. EMBO J 20:1042–1050
50. Kityk R, Kopp J, Sinning I, Mayer MP (2012) Structure and dynamics of the ATP-bound open conformation of Hsp70 chaperones. Mol Cell 48:863–874
51. Qi R, Sarbeng EB, Liu Q, Le KQ, Xu X, Xu H, Yang J, Wong JL, Vorvis C, Hendrickson WA, Zhou L, Liu Q (2013) Allosteric opening of the polypeptide-binding site when an Hsp70 binds ATP. Nat Struct Mol Biol 20:900–907
52. Vogel M, Bukau B, Mayer MP (2006) Allosteric regulation of Hsp70 chaperones by a proline switch. Mol Cell 21:359–367
53. Kityk R, Vogel M, Schlecht R, Bukau B, Mayer MP (2015) Pathways of allosteric regulation in Hsp70 chaperones. Nat Commun 6:8308

54. Sharma SK, De Los Rios P, Christen P, Lustig A, Goloubinoff P (2010) The kinetic parameters and energy cost of the Hsp70 chaperone as a polypeptide unfoldase. Nat Chem Biol 6:914–920
55. Rodriguez F, Arsène-Ploetze F, Rist W, Rüdiger S, Schneider-Mergener J, Mayer MP, Bukau B (2008) Molecular basis for regulation of the heat shock transcription factor sigma32 by the DnaK and DnaJ chaperones. Mol Cell 32:347–358
56. Lee JH, Zhang D, Hughes C, Okuno Y, Sekhar A, Cavagnero S (2015) Heterogeneous binding of the SH3 client protein to the DnaK molecular chaperone. Proc Natl Acad Sci U S A 112:E4206–E4215
57. De Los Rios P, De Los Rios P, Ben-Zvi A, Ben-Zvi A, Slutsky O, Slutsky O, Azem A, Azem A, Goloubinoff P, Goloubinoff P (2006) Hsp70 chaperones accelerate protein translocation and the unfolding of stable protein aggregates by entropic pulling. Proc Natl Acad Sci U S A 103:6166–6171
58. Goloubinoff P, Rios P (2007) The mechanism of Hsp70 chaperones:(entropic) pulling the models together. Trends Biochem Sci 32:372–380
59. Marcinowski M, Höller M, Feige MJ, Baerend D, Lamb DC, Buchner J (2011) Substrate discrimination of the chaperone BiP by autonomous and cochaperone-regulated conformational transitions. Nat Struct Mol Biol 18:150–158
60. Glover JR, Lindquist S (1998) Hsp104, Hsp70, and Hsp40: a novel chaperone system that rescues previously aggregated proteins. Cell 94:73–82
61. Goloubinoff P, Mogk A, Zvi AP, Tomoyasu T, Bukau B (1999) Sequential mechanism of solubilization and refolding of stable protein aggregates by a bichaperone network. Proc Natl Acad Sci U S A 96:13732–13737
62. Liberek K, Lewandowska A, Zietkiewicz S (2008) Chaperones in control of protein disaggregation. EMBO J 27:328–335
63. Rampelt H, Kirstein-Miles J, Nillegoda NB, Chi K, Scholz SR, Morimoto RI, Bukau B (2012) Metazoan Hsp70 machines use Hsp110 to power protein disaggregation. EMBO J 31:4221–4235
64. Nillegoda NB, Kirstein J, Szlachcic A, Berynskyy M, Stank A, Stengel F, Arnsburg K, Gao X, Scior A, Aebersold R, Guilbride DL, Wade RC, Morimoto RI, Mayer MP, Bukau B (2015) Crucial HSP70 co-chaperone complex unlocks metazoan protein disaggregation. Nature 524:247–251
65. Shorter J (2011) The mammalian disaggregase machinery: Hsp110 synergizes with Hsp70 and Hsp40 to catalyze protein disaggregation and reactivation in a cell-free system. PLoS One 6:e26319
66. Gao X, Carroni M, Nussbaum-Krammer C, Mogk A, Nillegoda NB, Szlachcic A, Guilbride DL, Saibil HR, Mayer MP, Bukau B (2015) Human Hsp70 disaggregase reverses Parkinson's-linked α-synuclein amyloid fibrils. Mol Cell 59:781–793
67. Liberek K, Galitski TP, Zylicz M, Georgopoulos C (1992) The DnaK chaperone modulates the heat shock response of Escherichia coli by binding to the sigma 32 transcription factor. Proc Natl Acad Sci U S A 89:3516–3520
68. Abravaya K, Myers MP, Murphy SP, Morimoto RI (1992) The human heat shock protein hsp70 interacts with HSF, the transcription factor that regulates heat shock gene expression. Genes Dev 6:1153–1164
69. Shi Y, Mosser DD, Morimoto RI (1998) Molecular chaperones as HSF1-specific transcriptional repressors. Genes Dev 12:654–666
70. Ron D, Walter P (2007) Signal integration in the endoplasmic reticulum unfolded protein response. Nat Rev Mol Cell Biol 8:519–529
71. Zylicz M (1993) The Escherichia coli chaperones involved in DNA replication. Philos Trans R Soc Lond B Biol Sci 339:271–277. discussion 277–278
72. Kim S-Y, Sharma S, Hoskins JR, Wickner S (2002) Interaction of the DnaK and DnaJ chaperone system with a native substrate, P1 RepA. J Biol Chem 277:44778–44783
73. Nakamura A, Wada C, Miki K (2007) Structural basis for regulation of bifunctional roles in replication initiator protein. Proc Natl Acad Sci U S A 104:18484–18489

74. Picard D (2002) Heat-shock protein 90, a chaperone for folding and regulation. Cell Mol Life Sci 59:1640–1648
75. Stankiewicz M, Mayer MP (2012) The universe of Hsp90. Biomol Concepts 3:79–97
76. Sousa R, Lafer EM (2015) The role of molecular chaperones in clathrin mediated vesicular trafficking. Front Mol Biosci 2:26
77. Xing Y, Böcking T, Wolf M, Grigorieff N, Kirchhausen T, Harrison SC (2010) Structure of clathrin coat with bound Hsc70 and auxilin: mechanism of Hsc70-facilitated disassembly. EMBO J 29:655–665
78. Böcking T, Aguet F, Harrison SC, Kirchhausen T (2011) Single-molecule analysis of a molecular disassemblase reveals the mechanism of Hsc70-driven clathrin uncoating. Nat Struct Mol Biol 18:295–301
79. Mayer MP (2005) Recruitment of Hsp70 chaperones: a crucial part of viral survival strategies. Rev Physiol Biochem Pharmacol 153:1–46
80. Sagara Y, Ishida C, Inoue Y, Shiraki H, Maeda Y (1998) 71-Kilodalton heat shock cognate protein acts as a cellular receptor for syncytium formation induced by human T-cell lymphotropic virus type 1. J Virol 72:535–541
81. Fang D, Haraguchi Y, Jinno A, Soda Y, Shimizu N, Hoshino H (1999) Heat shock cognate protein 70 is a cell fusion-enhancing factor but not an entry factor for human T-cell lymphotropic virus type I. Biochem Biophys Res Commun 261:357–363
82. Triantafilou K, Fradelizi D, Wilson K, Triantafilou M (2002) GRP78, a coreceptor for coxsackievirus A9, interacts with major histocompatibility complex class I molecules which mediate virus internalization. J Virol 76:633–643
83. Jindadamrongwech S, Thepparit C, Smith DR (2004) Identification of GRP 78 (BiP) as a liver cell expressed receptor element for dengue virus serotype 2. Arch Virol 149:915–927
84. Arias CF, Isa P, Guerrero CA, Méndez E, Zárate S, López T, Espinosa R, Romero P, López S (2002) Molecular biology of rotavirus cell entry. Arch Med Res 33:356–361
85. Chromy LR, Oltman A, Estes PA, Garcea RL (2006) Chaperone-mediated in vitro disassembly of polyoma- and papillomaviruses. J Virol 80:5086–5091
86. Ivanovic T, Agosto MA, Chandran K, Nibert ML (2007) A role for molecular chaperone Hsc70 in reovirus outer capsid disassembly. J Biol Chem 282:12210–12219
87. Zylicz M, Ang D, Liberek K, Georgopoulos C (1989) Initiation of lambda DNA replication with purified host- and bacteriophage-encoded proteins: the role of the dnaK, dnaJ and grpE heat shock proteins. EMBO J 8:1601–1608
88. Alfano C, McMacken R (1989) Heat shock protein-mediated disassembly of nucleoprotein structures is required for the initiation of bacteriophage lambda DNA replication. J Biol Chem 264:10709–10718
89. Liu JS, Kuo SR, Makhov AM, Cyr DM, Griffith JD, Broker TR, Chow LT (1998) Human Hsp70 and Hsp40 chaperone proteins facilitate human papillomavirus-11 E1 protein binding to the origin and stimulate cell-free DNA replication. J Biol Chem 273:30704–30712
90. Tanguy Le Gac N, Boehmer PE (2002) Activation of the herpes simplex virus type-1 origin-binding protein (UL9) by heat shock proteins. J Biol Chem 277:5660–5666
91. Baquero-Pérez B, Whitehouse A (2015) Hsp70 isoforms are essential for the formation of Kaposi's sarcoma-associated herpesvirus replication and transcription compartments. PLoS Pathog 11:e1005274
92. Song H, Moseley PL, Lowe SL, Ozbun MA (2010) Inducible heat shock protein 70 enhances HPV31 viral genome replication and virion production during the differentiation-dependent life cycle in human keratinocytes. Virus Res 147:113–122
93. Chromy LR, Pipas JM, Garcea RL (2003) Chaperone-mediated in vitro assembly of Polyomavirus capsids. Proc Natl Acad Sci U S A 100:10477–10482
94. Florin L, Becker KA, Sapp C, Lambert C, Sirma H, Müller M, Streeck RE, Sapp M (2004) Nuclear translocation of papillomavirus minor capsid protein L2 requires Hsc70. J Virol 78:5546–5553
95. Li PP, Itoh N, Watanabe M, Shi Y, Liu P, Yang H-J, Kasamatsu H (2009) Association of simian virus 40 vp1 with 70-kilodalton heat shock proteins and viral tumor antigens. J Virol 83:37–46

96. Kampinga HH, Craig EA (2010) The HSP70 chaperone machinery: J proteins as drivers of functional specificity. Nat Rev Mol Cell Biol 11:579–592
97. Gautschi M, Lilie H, Fünfschilling U, Mun A, Ross S, Lithgow T, Rücknagel P, Rospert S (2001) RAC, a stable ribosome-associated complex in yeast formed by the DnaK-DnaJ homologs Ssz1p and zuotin. Proc Natl Acad Sci U S A 98:3762–3767
98. Leidig C, Bange G, Kopp J, Amlacher S, Aravind A, Wickles S, Witte G, Hurt E, Beckmann R, Sinning I (2013) Structural characterization of a eukaryotic chaperone-the ribosome-associated complex. Nat Struct Mol Biol 20:23–28
99. Hundley HA, Walter W, Bairstow S, Craig EA (2005) Human Mpp11 J protein: ribosome-tethered molecular chaperones are ubiquitous. Science 308:1032–1034
100. Pais JE, Schilke B, Craig EA (2011) Reevaluation of the role of the Pam18:Pam16 interaction in translocation of proteins by the mitochondrial Hsp70-based import motor. Mol Biol Cell 22:4740–4749
101. Hagiwara M, Maegawa K-I, Suzuki M, Ushioda R, Araki K, Matsumoto Y, Hoseki J, Nagata K, Inaba K (2011) Structural basis of an ERAD pathway mediated by the ER-resident protein disulfide reductase ERdj5. Mol Cell 41:432–444
102. Lai CW, Otero JH, Hendershot LM, Snapp E (2012) ERdj4 protein is a soluble endoplasmic reticulum (ER) DnaJ family protein that interacts with ER-associated degradation machinery. J Biol Chem 287:7969–7978
103. Ushioda R, Hoseki J, Araki K, Jansen G, Thomas DY, Nagata K (2008) ERdj5 is required as a disulfide reductase for degradation of misfolded proteins in the ER. Science 321:569–572
104. Howarth JL, Kelly S, Keasey MP, Glover CPJ, Lee Y-B, Mitrophanous K, Chapple JP, Gallo JM, Cheetham ME, Uney JB (2007) Hsp40 molecules that target to the ubiquitin-proteasome system decrease inclusion formation in models of polyglutamine disease. Mol Ther 15:1100–1105
105. Sondermann H, Scheufler C, Schneider C, Höhfeld J, Hartl FU, Moarefi I (2001) Structure of a Bag/Hsc70 complex: convergent functional evolution of Hsp70 nucleotide exchange factors. Science 291:1553–1557
106. Shomura Y, Dragovic Z, Chang H-C, Tzvetkov N, Young JC, Brodsky JL, Guerriero V, Hartl FU, Bracher A (2005) Regulation of Hsp70 function by HspBP1: structural analysis reveals an alternate mechanism for Hsp70 nucleotide exchange. Mol Cell 17:367–379
107. Polier S, Dragovic Z, Hartl FU, Bracher A (2008) Structural basis for the cooperation of Hsp70 and Hsp110 chaperones in protein folding. Cell 133:1068–1079
108. Bracher A, Verghese J (2015) The nucleotide exchange factors of Hsp70 molecular chaperones. Front Mol Biosci 2:10
109. Kabbage M, Dickman MB (2008) The BAG proteins: a ubiquitous family of chaperone regulators. Cell Mol Life Sci 65:1390–1402
110. Lüders J, Demand J, Höhfeld J (2000) The ubiquitin-related BAG-1 provides a link between the molecular chaperones Hsc70/Hsp70 and the proteasome. J Biol Chem 275:4613–4617
111. Arndt V, Dick N, Tawo R, Dreiseidler M, Wenzel D, Hesse M, Fürst DO, Saftig P, Saint R, Fleischmann BK, Hoch M, Hohfeld J (2010) Chaperone-assisted selective autophagy is essential for muscle maintenance. Curr Biol 20:143–148
112. Gamerdinger M, Kaya AM, Wolfrum U, Clement AM, Behl C (2011) BAG3 mediates chaperone-based aggresome-targeting and selective autophagy of misfolded proteins. EMBO Rep 12:149–156
113. Behl C (2011) BAG3 and friends: co-chaperones in selective autophagy during aging and disease. Autophagy 7:795–798
114. Colón-Ramos DA, Irusta PM, Gan EC, Olson MR, Song J, Morimoto RI, Elliott RM, Lombard M, Hollingsworth R, Hardwick JM, Smith GK, Kornbluth S (2003) Inhibition of translation and induction of apoptosis by bunyaviral nonstructural proteins bearing sequence similarity to reaper. Mol Biol Cell 14:4162–4172
115. Minami R, Shimada M, Yokosawa H, Kawahara H (2007) Scythe regulates apoptosis through modulating ubiquitin-mediated proteolysis of the Xenopus elongation factor XEF1AO. Biochem J 405:495–501

116. Desmots F, Russell HR, Lee Y, Boyd K, McKinnon PJ (2005) The reaper-binding protein scythe modulates apoptosis and proliferation during mammalian development. Mol Cell Biol 25:10329–10337
117. Lee J-G, Ye Y (2013) Bag6/Bat3/Scythe: a novel chaperone activity with diverse regulatory functions in protein biogenesis and degradation. Bioessays 35:377–385
118. Liu Q, Hendrickson WA (2007) Insights into Hsp70 chaperone activity from a crystal structure of the yeast Hsp110 Sse1. Cell 131:106–120
119. Easton DP, Kaneko Y, Subjeck JR (2000) The hsp110 and Grp1 70 stress proteins: newly recognized relatives of the Hsp70s. Cell Stress Chaperones 5:276–290
120. Raviol H, Sadlish H, Rodriguez F, Mayer MP, Bukau B (2006) Chaperone network in the yeast cytosol: Hsp110 is revealed as an Hsp70 nucleotide exchange factor. EMBO J 25:2510–2518
121. Dragovic Z, Broadley SA, Shomura Y, Bracher A, Hartl FU (2006) Molecular chaperones of the Hsp110 family act as nucleotide exchange factors of Hsp70s. EMBO J 25:2519–2528
122. Steel GJ, Fullerton DM, Tyson JR, Stirling CJ (2004) Coordinated activation of Hsp70 chaperones. Science 303:98–101
123. Goeckeler JL, Petruso AP, Aguirre J, Clement CC, Chiosis G, Brodsky JL (2008) The yeast Hsp110, Sse1p, exhibits high-affinity peptide binding. FEBS Lett 582:2393–2396
124. Xu X, Sarbeng EB, Vorvis C, Kumar DP, Zhou L, Liu Q (2012) Unique peptide substrate binding properties of 110-kDa heat-shock protein (Hsp110) determine its distinct chaperone activity. J Biol Chem 287:5661–5672
125. Li Z, Hartl FU, Bracher A (2013) Structure and function of Hip, an attenuator of the Hsp70 chaperone cycle. Nat Struct Mol Biol 20:929–935
126. Mayer MP, Le Breton L (2015) Hsp90: breaking the symmetry. Mol Cell 58:8–20
127. Röhl A, Rohrberg J, Buchner J (2013) The chaperone Hsp90: changing partners for demanding clients. Trends Biochem Sci 38:253–262
128. Morishima Y, Kanelakis KC, Silverstein AM, Dittmar KD, Estrada L, Pratt WB (2000) The Hsp organizer protein hop enhances the rate of but is not essential for glucocorticoid receptor folding by the multiprotein Hsp90-based chaperone system. J Biol Chem 275:6894–6900
129. Kirschke E, Goswami D, Southworth D, Griffin PR, Agard DA (2014) Glucocorticoid receptor function regulated by coordinated action of the hsp90 and hsp70 chaperone cycles. Cell 157:1685–1697
130. Karagöz GE, Duarte AMS, Akoury E, Ippel H, Biernat J, Morán Luengo T, Radli M, Didenko T, Nordhues BA, Veprintsev DB, Dickey CA, Mandelkow E, Zweckstetter M, Boelens R, Madl T, Rüdiger SGD (2014) Hsp90-Tau complex reveals molecular basis for specificity in chaperone action. Cell 156:963–974
131. Meacham GC, Patterson C, Zhang W, Younger JM, Cyr DM (2001) The Hsc70 co-chaperone CHIP targets immature CFTR for proteasomal degradation. Nat Cell Biol 3:100–105
132. Connell P, Ballinger CA, Jiang J, Wu Y, Thompson LJ, Höhfeld J, Patterson C (2001) The co-chaperone CHIP regulates protein triage decisions mediated by heat-shock proteins. Nat Cell Biol 3:93–96
133. Höhfeld J, Cyr DM, Patterson C (2001) From the cradle to the grave: molecular chaperones that may choose between folding and degradation. EMBO Rep 2:885–890
134. Stankiewicz M, Nikolay R, Rybin V, Mayer MP (2010) CHIP participates in protein triage decisions by preferentially ubiquitinating Hsp70-bound substrates. FEBS J 277:3353–3367
135. Calloni G, Chen T, Schermann SM, Chang H-C, Genevaux P, Agostini F, Tartaglia GG, Hayer-Hartl M, Hartl FU (2012) DnaK functions as a central hub in the E. coli chaperone network. Cell Rep 1:251–264
136. Marcinowski M, Rosam M, Seitz C, Elferich J, Behnke J, Bello C, Feige MJ, Becker CFW, Antes I, Buchner J (2013) Conformational selection in substrate recognition by Hsp70 chaperones. J Mol Biol 425:466–474

# Chapter 2
# Molecular Chaperone Inhibitors

**Michael A. Moses, Abbey D. Zuehlke, and Len Neckers**

**Abstract** Hsp70 and Hsp90 are molecular chaperones (heat shock proteins) that facilitate client protein maturation, stabilization of aggregation-prone proteins, quality control of misfolded proteins and maintenance of proteins in an activation-competent conformation. In general, these Hsps are part of the cellular proteostasis network that functions in normal and disease states to maintain protein homeostasis. Recent data suggest a role for certain components of the proteostasis network (e.g., the proteasome) in various aspects of immune responses, and lately molecular chaperones have also been suggested to play a role in immunity, although the exact nature of their function remains somewhat controversial. Given the growing importance of Hsp90 and Hsp70 in a number of different diseases, including cancer and neurodegenerative maladies, as well as their role in contributing to protein homeostasis in health and disease, pharmacologic targeting of Hsp70, Hsp90 and their respective co-chaperones remains an area of intense investigation, although the impact of Hsp inhibition on immune cells and systems remains poorly understood.

**Keywords** Heat shock protein 90 · Heat shock protein 70 · Proteostasis · Chaperone inhibitors · Co-chaperone inhibitors

## 2.1 The Hsp70 and Hsp90 Chaperone Cycle

Hsp70 and Hsp90 are chaperones that facilitate protein maturation, stabilization of aggregation-prone proteins, quality control of misfolded proteins, and maintenance of proteins in an activation-competent conformation. Proteins that rely on Hsp70 and Hsp90 for function are shuttled from Hsp70 to Hsp90 utilizing a co-chaperone-assisted cycle. Co-chaperones play a role in client transfer, ATPase regulation and conformational dynamics of chaperones. Given the importance of these chaperones

M. A. Moses · A. D. Zuehlke · L. Neckers (✉)
Urologic Oncology Branch, Center for Cancer Research, National Cancer Institute, Bethesda, MD, USA
e-mail: neckers@nih.gov



R. J. Binder, P. K. Srivastava (eds.), *Heat Shock Proteins in the Immune System*,
https://doi.org/10.1007/978-3-319-69042-1_2

in a number of different diseases, including cancer and neurodegenerative maladies, targeting Hsp70, Hsp90, and their respective co-chaperones remains an area of intense investigation, although the impact of such strategies on immune cells and systems remains less well understood [1, 2].

## 2.2 Hsp90 Inhibitors and Their Binding Sites

### *2.2.1 Hsp90 N-Terminal Inhibitors*

Hsp90 inhibition gained interest following the discovery of Hsp90 as a target for the natural products geldanamycin (GA) and radicicol (RD) [3–5]. Drugging Hsp90 through the use of GA resulted in reduced cancer cell line growth and caused oncogenic protein depletion [6]. Hsp90 exists as a dimeric protein with each monomer containing an N-terminal domain (NTD) that binds ATP, a middle domain that interacts with co-chaperones and clients (and is required for ATP hydrolysis), and a C-terminal dimerization domain. These regions undergo substantial ATP-influenced conformational changes in order for Hsp90 to properly chaperone its clientele. The absence of bound nucleotide results in Hsp90 preferring an open conformation (NTDs not dimerized) with its C-terminal domains dimerized. Nucleotide interaction results in transient dimerization of the NTDs and strengthens Hsp90-client interactions. Following ATP hydrolysis, Hsp90 proceeds back to the open conformation. Crystal structure analysis determined the binding site for GA and RD to be within the Hsp90 N-terminal ATP-binding domain [7–9]. Although these compounds efficiently disrupted Hsp90 function, they were not useful for clinical application due to their toxicity and low stability.

The selectivity of GA and RD toward Hsp90 is due to the fact that the N-terminal ATP-binding pocket of Hsp90 contains the unique Bergerat fold geometry found only within the select GHKL subgroup of ATPases [7, 10]. This has allowed development of less toxic, highly specific, and more stable Hsp90 inhibitors that mimic GA and RD interaction within the ATP-binding pocket.

### *2.2.2 Benzoquinone Ansamycin Inhibitors*

GA is a benzoquinone ansamycin antibiotic isolated from *Streptomyces hygroscopicus*. Once bound to Hsp90, GA adopts a folded position with the planes of the benzoquinone and the ansamycin macrocycle close to parallel [9]. The benzoquinone group binds near the entrance of the pocket, and the ansamycin ring, which resembles a five amino acid polypeptide, inserts into the pocket [8]. The most notable site of interaction is within the deepest contacts of the pocket at the

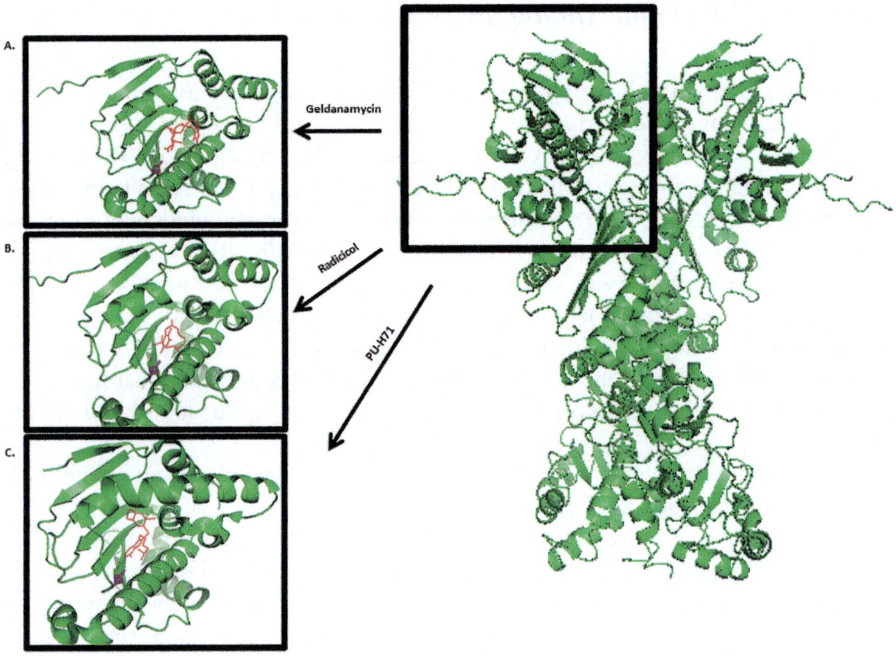


**Fig. 2.1** Three N-terminal domain (NTD) binding Hsp90 inhibitors utilize different contacts within the ATP-binding pocket and differentially affect NTD subdomains. Crystal structure of the Hsp90 dimer is shown on right. The NTD of one Hsp90 protomer (within black box) is shown in the left panels bound to, respectively, the benzoquinone ansamycin geldanamycin (**a**), the macrocyclic lactone radicicol (**b**), or the purine analog PU-H71 (**c**). In each case, drugs are shown in red. Refer to text for details

highly conserved amino acid Asp93 of Hsp90. This is because the region in which Asp93 is located is otherwise mostly hydrophobic [8]. Mutation of this position disrupts Hsp90 function and inhibits nucleotide binding [11, 12]. Although GA is effective at inhibiting Hsp90, the bioreduction of the semiquinone radical creates a toxic superoxide radical [13].

A compound closely related to GA, 17-allylamino-17-demethoxy geldanamycin (17-AAG), was the first Hsp90 inhibitor to enter into clinical trials. This compound replaced the 17-methoxy moiety of GA with a 17-alkylamino group to reduce toxicity [14]. Despite displaying anticancer efficacy in Phase I clinical trials, especially in combination with trastuzumab in HER-2 positive breast cancer patients, 17-AAG development was discontinued due to poor water solubility and dose-limiting hepatotoxicity [15]. Other benzoquinone ansamycin analogs that have been evaluated in humans include 17-DMAG, IPI-504, and 17-AG. However, there are currently no benzoquinone ansamycin compounds in clinical trials. The binding pose of the benzoquinone ansamycins to Hsp90 is shown in Fig. 2.1a.

### 2.2.3 Radicicol and Analogous Inhibitors

RD, another natural product Hsp90 inhibitor, is a macrocyclic lactone antibiotic that was isolated from *Monosporium bonorden* in 1953 [16]. In 1998, RD was found to compete with GA for the Hsp90 ATP interaction site [5]. Similar to GA, RD binding mimics the ADP-bound conformation of Hsp90 and interacts with Asp93 in a manner similar to GA (Fig. 2.1b). Compared to GA, however, RD binds in a different orientation and has greater affinity for the ATP pocket [9]. RD is oriented in the opposite direction of GA with the aromatic ring directed toward the bottom of the pocket and the macrocycle facing toward and making contacts with the top of the pocket. RD also adopts a folded conformation, but it is less dramatic than that of GA, with the macrocycle and aromatic ring approximately perpendicular instead of parallel [9]. RD does not, however, display antitumor activity due to its rapid metabolism in vivo [17].

To increase stability, the 2′ ketone of radicicol was converted to an oxime [18]. This led to the production of KF25706, which is metabolically stable and displays potency against several human cancer cell lines and rodent xenograft models [18]. The complexity of this compound, however, made it difficult for large-scale production. The resorcinol moiety of RD behaves in a similar manner as the adenine ring of ATP and is required for Hsp90 inhibition. Utilizing the resorcinol ring, several inhibitors have been created and have undergone clinical evaluation. A resorcinol triazole compound created by Synta Pharmaceuticals (STA-9090, ganetespib) has been shown to have high affinity for Hsp90 and to inhibit its activity at concentrations as low as 10 nM. Ganetespib also displays increased tumor penetration with low toxicity [17].

Workman and co-workers screened a library of 56,000 compounds and identified CCT018159, which contains the resorcinol-anchoring unit of radicicol [19]. Further development of CCT018159 resulted in the creation of the resorcinylic isoxazole amide NVP-AUY922/VER52296 (Novartis/Vernalis). Additional resorcinol analogs include KW-2478 (Kyowa Hakko Kirin Pharma) and AT13387 (Astex Pharmaceuticals). Currently, ganetespib is involved in several clinical trials, as are KW-2478 and AT13387. NVP-AUY922 is no longer being clinically evaluated due to failure to show clinically meaningful responses at the maximum tolerated dose [17].

### 2.2.4 Additional Hsp90 Second-Generation Inhibitors

Utilizing knowledge of the natural product inhibitors' interactions with the Hsp90 ATP pocket resulted in the development of additional synthetic compounds with increased potency and reduced toxicity. Purine-based compounds were designed to mimic the folded structure adopted by GA and RD when bound to Hsp90. These compounds contain an adenine group, a CH2 or S linker, and a right side aryl group. The adenine binds the pocket in a similar manner to ATP and maintains the direct hydrogen bond from N6 to Asp93 on Hsp90. The first inhibitor created was PU3 [20, 21]. Although this compound showed efficacy in oncogenic cell lines, it was not as potent as 17-AAG. Further modification of PU3 to improve interaction with Hsp90 resulted in the synthesis of PU-H71. This compound shows higher activity toward oncogenic cells and is a potent inhibitor of Hsp90 activity [22, 23]. PU-H71 contains an N9 alkane off of the adenine group. This amine moiety protrudes from the ATP-binding cavity out into the solvent, and, as it does not interact directly with the protein, it is amenable to further modifications that may improve its pharmacological properties (Fig. 2.1c) [24]. Currently, PU-H71 is being evaluated in patients with advanced malignancies. Other purine scaffolds have been created including CNF2024/BIIB021 (Biogen Idec), MPC-3100 (Myriad Pharmaceuticals Inc.), and Debio 0932 (Curis). CNF2024/BIIB021 and Debio 0932 have both been evaluated in Phase I and II clinical trials.

Benzamide compounds represent another class of Hsp90 inhibitors. These drugs utilize their benzamide group to mimic adenine, with the amide group forming hydrogen bonds with Asp93 and Thr184 [25]. Benzamide Hsp90 inhibitors include SNX-5422, developed by Serenex and acquired by Pfizer, which is a prodrug of the active metabolite SNX2112. SNX-5422, now owned by Esanex, remains under clinical evaluation. Another benzamide compound, created by Taiho Pharmaceutical, Co. Ltd., is the 4-(1*H*-pyrazolo[3,4-*b*]pyridine-1-yl)benzamide TAS-116 [26]. TAS-116 is currently being evaluated in recently initiated clinical trials.

### *2.2.5 Consequences of Different N-Terminal Inhibitors for Hsp90 Specificity and Conformation*

From protein crystallography, it is clear that these chemically diverse NTD-interacting Hsp90 inhibitors bind to the ATP pocket and share many of the same contacts. However, there are subtle differences in these drugs' interactions with Hsp90 and their sensitivity to various identified posttranslational modifications. These properties result in subtle changes in Hsp90 conformation upon drug binding [27]. Figure 2.1 displays some of the different NTD inhibitor interactions with Hsp90. Based on these data, it is likely that a better understanding of each inhibitor's binding preferences may allow for the creation of inhibitors with greater cellular specificity and/or potency.

### *2.2.6 Hsp90 C-Terminal Inhibitors*

Utilizing nucleotide affinity cleavage, a second ATP-binding pocket was discovered within the C-terminus of Hsp90 [28]. The natural product novobiocin was the first C-terminal inhibitor identified. The site of novobiocin interaction is proximal to the C-terminal dimerization domain, and novobiocin binds in a bent ADP-like state. This C-terminal nucleotide pocket does not interact with GA or RD and novobiocin does not interact with the Hsp90 NTD [29]. Novobiocin structure contains three features: a benzamide side chain, a coumarin core, and a noviose sugar. Utilizing the structure of novobiocin, Blagg and colleagues synthesized the compound A4 and its analogs. These compounds were coumarin-modified ring systems that mimic adenine and guanine with additional strategically placed hydrogen bond acceptors and donors to fit the C-terminal nucleotide pocket with greater affinity and higher specificity [30]. The most potent novobiocin analog created to date is KU-174 [31]. This compound has shown efficacy in several cancer cell lines and promotes the degradation of Hsp90 clients without induction of the heat shock response (HSR)—a transcriptional program that responds to environmental stress to promote survival in both normal and transformed cells [32]. An additional C-terminal inhibitor is epigallocatechin-3-gallate (EGCG), the most abundant catechin in green tea. This compound has reported anticancer activity and was found to interact with the same region in Hsp90 as novobiocin (amino acids 538–728) [33–35]. Other less-specific C-terminal inhibitors include the platinum-containing chemotherapeutic agent, cisplatin, and the microtubule stabilizer, Taxol [36]. To date, none of the novobiocin-derived C-terminal inhibitors have been evaluated in the clinic. However, the ability of these compounds to inhibit Hsp90 without inducing the cytoprotective HSR makes their continued development worthwhile.

## 2.3 Grp94

Grp94 is an endoplasmic reticulum (ER)-restricted Hsp90 paralog. Like the cytosolic Hsp90 paralogs, nucleotide binding is important for Grp94's chaperone activity. Grp94 helps to buffer proteotoxic stress in the ER and regulates the stability of membrane-associated and membrane-secreted proteins, including immunoglobulins, Toll-like receptors, integrins, and growth factors [37]. As such, Grp94 is likely to play a role in immune function and diseases associated with the secretory pathway.

### 2.3.1 *Allosteric Inhibitors of Grp94*

Due to the structural similarities of the ATP-binding pocket among Hsp90 paralogs, all NTD Hsp90 inhibitors not only target cytosolic Hsp90α/β but Grp94 (and TRAP-1, the mitochondrial Hsp90 paralog) as well, making it impossible to decipher the individual contribution(s) of these paralogs to different disease states [6, 38]. However, unlike the other Hsp90 paralogs, Grp94's ATP-binding pocket contains a QEDGQ amino acid insertion that creates a unique secondary hydrophobic binding cleft. This structural distinction is the basis for the development of Grp94-selective inhibitors.

Combining library screening with computational modeling, the Chiosis laboratory examined purine scaffolds that possessed a higher affinity for Grp94 compared to the other Hsp90 paralogs. One such compound, PU-H54, was demonstrated to be highly selective for Grp94, as evidenced by decreased IGF-II secretion and membrane trafficking of Toll-like receptors. Furthermore, the use of PU-H54 revealed the specificity of Grp94 for membrane-associated HER2 in high HER2-expressing breast cancer cells, suggesting a role for support of the malignant phenotype by Grp94 in a tumor-selective manner [39].

The Blagg laboratory has also synthesized selective Grp94 inhibitors. The co-crystal structure of the pan-Hsp90 inhibitor, radamide, with yeast cytosolic Hsp90 and canine Grp94 identified differential binding interactions with these chaperones [40]. Optimization of structure-activity relationships led to the development of radamide analogs with higher affinity for Grp94, including BnIm and KUNG29 (compound 40), that interact with Phe199 and Tyr200 in Grp94's secondary pocket. Cell-based assays determined that these Grp94 inhibitors also block IGF-II and Toll-like receptor trafficking [40–42]. Furthermore, they display anti-proliferative activity against a multiple myeloma (immunoglobulin secreting) cell line and anti-migratory behavior in a breast cancer model [41], consistent with Grp94's role in the maturation and trafficking of proteins involved in protein secretion and metastasis.

## 2.4 Hsp90 Co-Chaperone Inhibitors

Hsp90 function relies on an ordered progression through a number of nucleotide-influenced conformations. Throughout its ATPase cycle, helper co-chaperone proteins bind and release Hsp90 in order to assist in processes such as conformational dynamics and nucleotide and client interactions. Although the functional importance of all the co-chaperones has not been fully characterized, studies focused on individual co-chaperones demonstrate that they each play distinct roles with respect to Hsp90 regulation. Co-chaperone proteins are also uniquely regulated in different illnesses, thus identifying them as molecular targets of potential interest for pharmaceutical development. $p50^{Cdc37}$ is a co-chaperone that binds the open (NTD undimerized) conformation of Hsp90 and is known for its role in recruiting kinase clients to the Hsp90 chaperone machinery, as si-RNA knockdown of $p50^{Cdc37}$ results in client kinase degradation [43]. Furthermore, ATP-competitive kinase inhibitors such as vemurafenib and lapatinib have been found to disrupt kinase-$p50^{Cdc37}$ interaction, which also results in degradation of the oncogenic kinases B-Raf and ErbB2, respectively [44]. Another client recruiting co-chaperone, Hop, binds the open conformation of Hsp90 and bridges Hsp70 to Hsp90 for client loading. Disruption of Hop-Hsp90 interaction leads to the proteasome-mediated degradation of a diverse set of chaperone clients and causes cell cycle arrest, inhibition of cell adhesion, and apoptosis [45].

Following ATP interaction, Hsp90 proceeds into a closed (NTD dimerized) conformation. While in this conformation, Hsp90 interacts with immunophilin proteins at its C-terminus. These immunophilins are most well studied for their impact on steroid hormone regulation. FKBP51 is an immunophilin made up of two FKBP-like domains, which contain its peptidyl-prolyl isomerase activity, as well as a tetratricopeptide repeat (TPR) domain for its interaction with the Hsp90 C-terminal MEEVD motif. Although FKBP51 binds several nonselective inhibitors including cyclosporin A, rapamycin, and FK506, a recent study identified ligands that selectively interact with and inhibit FKBP51 [46, 47]. The use of these ligands in mice resulted in improved endocrine feedback and stress-coping behavior, suggesting a new paradigm for antidepressant development [47].

P23 interacts with the N-terminal domain of Hsp90 while it is in complex with the immunophilin proteins and stabilizes the closed conformation. As Hsp90 interacts strongly with clients while in the closed conformation, inhibition of p23 results in client instability. The natural product celastrol inhibits p23 by altering its three-dimensional structure, leading to rapid formation of amyloid fibrils [48]. Another natural product, gedunin, was found to bind directly to p23 and to inactivate its function, as well as to disrupt its interaction with Hsp90. This inhibition resulted in destabilized nuclear receptors, with no impact on Hsp90-dependent kinases, and led to cancer cell death via apoptosis [49]. Further analysis of co-chaperone interaction with and regulation of Hsp90 may allow for alternative methods of inhibiting Hsp90 function.

## 2.5 Hsp70/Hsc70

Simply, there are two main forms of Hsp70, one that is constitutively expressed (Hsc70) and one that is stress induced (Hsp70), although each main form is comprised of multiple isoforms. Hsp70 (unless otherwise noted "Hsp70" will be used to refer to all family members) consists of two main domains: a nucleotide-binding domain (NBD), responsible for binding ATP, and a substrate-binding domain (SBD), which binds hydrophobic regions of immature or misfolded proteins. The NBD is further divided into four subdomains IA, IIA, IB, and IIB, which encompass a deep ATP-binding pocket. Residues implicated in the binding of nucleotide to Hsc70 include Glu268, Lys271, Arg272, Ser275, and Arg342 in subdomain IIB and Thr13, The14, and Tyr15 in subdomain IA [50, 51]. Hsp70 interaction with co-chaperones such as J-domain proteins (Hsp40s/DnaJs) and nucleotide exchange factors (Bag family members) influence its ATPase activity and provide substrate specificity. Because Hsp70 regulates numerous oncogenic client proteins (many of which rely on Hsp90 for stability) and chaperones other client proteins involved in neurodegenerative disease and viral infections, Hsp70 is an attractive therapeutic target for a number of pathologies [52]. Evidence suggests that Hsp70 is druggable and inhibitors have been identified that bind to either the NBD or SBD. Herein we discuss three main types of Hsp70 inhibition: ATP competitors, allosteric inhibitors, and peptide mimetics.

## 2.6 Hsp70/Hsc70 Inhibitors and Their Binding Sites

### *2.6.1 ATP-Competitive Inhibitors of Hsp70/Hsc70*

The strong binding affinity of ATP for Hsp70 has made it difficult to develop competitive inhibitors. However, Vernalis took advantage of structure-based design and utilized the adenosine fragment of ATP as a suitable starting point for further development and optimization of agents targeting the Hsp70 NBD. Because an X-ray crystal structure of full-length human Hsp70 has yet to be solved and crystal structures of isolated Hsp70 domains are not suitable for drug binding studies, Vernalis took advantage of the crystal structure of Hsc70 (whose sequence and structure are similar to that of Hsp70) complexed with the nucleotide exchange factor Bag1 to identify a hit molecule binding to the Hsc70/Hsp70 ATP-binding cleft [51]. Following sequential rounds of structure-guided synthesis, VER-155008 was developed as a pan-Hsp70 ATP-competitive inhibitor. VER-155008 competes with ATP for binding to the Hsp70 NBD, contacting similar residues as ATP (Tyr15, Arg272), and possesses biological activity in a variety of model systems [53]. It inhibits the proliferation and migration of cancer cells [54] and has been used to confirm a role for Hsp70 in pathogenicity and virulence [55].

In a cell-based screen to identify imidazoles with apoptosis-inducing activity, Williams and colleagues discovered apoptozole as an Hsc70/Hsp70 interactor [56]. Further elucidation of its mechanism of action determined that apoptozole binds to the ATPase domain of Hsp70, interacting with Ser275 and Arg272 in the ATP-binding pocket, and inhibits its ATPase activity [57, 58]. The pro-apoptotic effects of apoptozole are attributed to its ability to block the interaction of Hsp70 with apoptotic protease-activating factor-1 (APAF-1), a known Hsp70 client protein, thereby promoting caspase-dependent cell death in vitro and in vivo [58].

### *2.6.2 Allosteric Inhibitors of Hsp70/Hsc70*

One of the first reports of an Hsp70 inhibitor identified MKT-077, a cationic rhodocyanine isolated from a screen of compounds that exhibited effects on cancer cell viability [59, 60]. MKT-077 was first shown to bind Hsc70 [61] and the mitochondrial Hsp70 family member mortalin [62]. However, it wasn't until 2011 when the Gestwicki laboratory relied on NMR to show MKT-077 interacts with ADP-bound Hsc70 by inserting itself in a negatively charged allosteric pocket near the NBD (although it does not compete with nucleotide for binding). Interestingly, the residues at this site (Glu175, Asp199, Asp206) are conserved among major Hsp70 family members and appear to be important in ATP hydrolysis [63]. The ability of MKT-077 to lock Hsp70 in an ADP-bound conformation, where interaction with clients is strongest, likely halts the chaperone cycle in place and allows for interaction with Hsp90 and with the E3-ubiquitin ligase co-chaperone CHIP (C-terminal

Hsp-interacting protein). CHIP/Hsp70 complexes are thought to function in protein triage to clear cells of terminally misfolded proteins. Although MKT-077 is efficacious in some cancer models [64], it was found to be rapidly metabolized in mouse liver microsomes, limiting its clinical utility [65].

Identification of the binding pocket and orientation of MKT-077 led to the generation of YM-01, an MKT-077 analog. Like MKT-077, YM-01 locks Hsc70/Hsp70 in an ADP-bound conformation where interaction with substrates is strongest, thus delaying their dissociation from the chaperone complex and promoting their ubiquitination by CHIP. Notably, both MKT-077 and YM-01 are not toxic to normal cells [66–68], suggesting these compounds and their analogs may only be efficacious in cells addicted to Hsp70, similar to results with Hsp90 inhibitors [1]. YM-01 has been evaluated in preclinical models where Hsp70 plays a role in buffering proteotoxic stress. In breast cancer cells, YM-01 affects a number of signaling nodes controlled by Hsp70 (FoxM1, survivin, p21, HuR, Src, Hif1) in vitro and in vivo [69]. Studies in this model system revealed the importance of the Hsp70 nucleotide exchange factor, Bag3. Treatment of breast cancer cells with YM-01 not only locked Hsp70 in an ADP-bound conformation but concomitantly weakened its interaction with nucleotide exchange factor Bag3, thus preventing substrate release and stalling the chaperone cycle [69]. YM-01 has also been shown to have activity in neurodegenerative disorders. In spinobulbar muscular atrophy, characterized by the accumulation of toxic protein aggregates, YM-01 promotes the clearance of toxic polyglutamine-expanded androgen receptor aggregates to alleviate neurotoxicity [70]. Others have used YM-01 in Alzheimer's disease models to reduce tau levels and restore synaptic function [68]. Notably, it appears that the critical reliance of cells on Hsp70 for protein triage decisions identifies neurodegenerative disorders as a unique group of diseases in which targeting of Hsp70 may prove to be beneficial [71].

Further optimization of the MKT-077 and YM-01 scaffold led to the synthesis of JG98 [72]. JG98 occupies the same allosteric site that binds MKT-077 but has markedly increased stability and is significantly more potent in inhibiting breast cancer proliferation in vitro and xenograft growth in vivo compared to MKT-077 [69, 73]. Collectively, these results support the hypothesis that Hsp70-Bag3 regulates a number of oncogenic signaling pathways. In addition, the use of these inhibitors has validated the role of Hsp70 in viral protein homeostasis and replication of Dengue virus and other flaviviruses including yellow fever, West Nile virus, and Japanese encephalitis virus [74]. As such, viral infections are likely an additional indication for which these inhibitors may be efficacious.

In a screen of purine-like compounds that discriminate between Hsp70 and Hsc70, the Haystead laboratory identified HS-72. HS-72 is a selective allosteric inhibitor of Hsp70 that does not inhibit Hsp70-mediated ATP hydrolysis but, instead, induces a conformational change reducing ATP affinity. Site-directed mutagenesis suggests HS-72 may bind at or near cysteine 306, a non-conserved residue among Hsp70 family members that potentially allows its Hsp70-selective activity. HS-72 possesses cellular activity as it inhibits cancer cell proliferation and leads to protein aggregation in vitro. HS-72 is also effective in the MMTV-*neu* breast cancer mouse model [75] and displays anti-dengue virus activity [76].

The Chiosis laboratory utilized computational modeling to identify druggable allosteric sites in Hsp70. YK5, whose chemical structure is based upon a 2,5′-thiopyrimidine scaffold, targets a cleft region outside the Hsp70 ATP/ADP binding domain (around subdomains IB and IIB) and forms a covalent bond with a reactive cysteine (cysteine 267) inside this pocket. Modeling also suggests YK5 interacts with Leu237, Val238, Asp234, Arg 264, and Glu268. In-cell binding assays revealed YK5 binds both Hsc70 and Hsp70 and inhibits the Hsp70-dependent folding of an artificial substrate. Consistent with these effects, YK5 promotes the degradation of Her2, Raf-1, and Akt, oncogenic kinases that utilize the Hsp70 chaperone complex for stability prior to Hsp90-associated maturation [77].

Other strategies to target Hsp70 involve manipulating its interaction within higher-order multi-chaperone complexes. Gestwicki and colleagues screened a library of compounds and identified the flavonoid myricetin as binding to an allosteric pocket on bacterial Hsp70 (DnaK) adjacent to the NBD. By interacting with subdomains IB and IIB of DnaK, myricetin precludes DnaJ (bacterial Hsp40) from interacting with and stimulating DnaK's ATPase activity [78]. The resulting structural change also prevents the binding and chaperoning of substrates by DnaK, a property which may underlie myricetin's anticancer activity.

The Gestwicki laboratory has also identified additional compounds that stimulate or inhibit Hsp70 function in the presence of J-domain (Hsp40) proteins. Based on an earlier observation that large T-antigen (which has a J-domain) stimulation of Hsp70 ATPase activity could be inhibited by the dihydropyrimidine MAL3-101 [79], similar structure-activity guided synthesis led to identification of the dihydropyrimidine compound 115-7c [80]. 115-7c binds to subunit IIA of DnaK at a site partially overlapping with J-domain responsive residues and acts as an artificial co-chaperone, stimulating Hsp70 ATPase activity (in the presence of ATP or ADP). Residues in the adjacent IA subdomain are also responsive to 115-7c promotion of ATPase activity. Interestingly, 115-7c binds with better affinity to DnaK in the presence of DnaJ, which may be attributed to a more favorable binding pose in the DnaK-DnaJ complex. Compound 115-7c was also converted from activator to inhibitor (116-9e) by addition of a bulky diphenyl group to create steric clash with DnaK, thereby inhibiting DnaK-DnaJ interaction and DnaK ATPase activity [80].

2-Phenylethynesulfonamide (PES) and its derivatives (PES-Cl, PET-16) are the only inhibitors shown to bind to the Hsp70 substrate-binding domain. PES was identified from a screen of small molecules that activated p53-mediated apoptosis [81]. Probing cell lysate with biotinylated PES led to the identification of Hsc70 and Hsp70 as its targets [82]. The use of truncation mutants and co-crystal structures revealed that PES and its analog PET-16 bind to the Hsp70 SBD in the ADP-bound form, making contact with residues L392, P396, L399, G482, A503, and S504 of Hsp70 [82–85]. Efficacious in cell culture and in a Myc-induced lymphoma mouse model, these inhibitors block Hsp70 interactions with client proteins and co-chaperones (CHIP, Bag-1, Hsp40). As a result, client proteins accumulate in detergent-insoluble fractions, and both the ubiquitin-proteasome and lysosome-autophagy cellular protein clearance systems are disrupted [86].

### 2.6.3 Peptide Mimetics Targeting Hsp70/Hsc70

As mentioned, Hsp70 interacts with Bax and APAF-1 to suppress their activity. Other pro-apoptotic interactors of Hsp70 include apoptosis-inducing factor (AIF), which binds to the Hsp70 SBD and is sequestered by Hsp70. By perturbing this interaction with ADD70, a peptide mimetic of AIF, cancer cells are sensitized to a variety of pro-death signals [87]. Furthermore, ADD70 treatment potentiated the infiltration of $CD8^+$ T cells in tumors in vivo, pointing to a role of Hsp70 in regulating anticancer immunity [88]. The discovery of peptide aptamers that bind Hsp70 has further validated the concept that antagonism of Hsp70 sensitizes cancer cells to anticancer agents in vitro and in vivo while affecting host immune function. Treatment of B16F10 tumor-bearing mice with an aptamer targeting the Hsp70 ATP-binding domain triggered a host antitumor response associated with increased infiltration of macrophages and T lymphocytes, confirming the importance of Hsp70 in anticancer immunity [89].

## 2.7 Hsp40

Hsp40 (a family of DnaJ/J-domain proteins) functions to prevent aggregation of nascent polypeptides. The chaperone is also responsible for delivering clients to Hsp70 and for stimulating Hsp70 ATPase activity. Although pharmacologic targeting of Hsp40 has not been extensively explored, the fact that Hsp40 binds numerous oncoclients and has been shown to play a role in the viral life cycle suggests that it may be a promising molecular target in cancer and viral diseases [90]. From a screen of drug-like compounds, a group of phenoxy-*N*-arylacetamides was identified for their ability to inhibit protein refolding in a reticulocyte lysate-mediated luciferase refolding assay. Using a more defined system with purified Hsp70 and DnaJ proteins, it was shown that this class of compounds binds directly to DnaJ and inhibits Hsp70/DnaJ-mediated luciferase refolding [91].

## 2.8 Small Hsps-Hsp27

Hsp27 is as a stress-inducible, HSF-1-regulated chaperone protein that functions in an ATP-independent manner, making it difficult to target with small molecules. Instead, Hsp27 activity is regulated by phosphorylation and undergoes a high degree of oligomerization (like other small Hsps): while small oligomers foster degradation of client proteins through ubiquitin-mediated proteasome degradation, large oligomers act to stabilize misfolded proteins and to prevent aggregation. Hsp27 has been reported to be cytoprotective, preventing apoptosis during proteotoxic stress, and thus it may support the malignant phenotype [92].

### *2.8.1 Antisense Oligonucleotides Targeting Hsp27*

OGX-427 is a second-generation antisense oligonucleotide targeting Hsp27 mRNA. Hsp27 has been shown to facilitate androgen receptor (AR) transcriptional activity in prostate cancer cells, highlighting a novel role for this Hsp [93]. Treatment of prostate cancer cells with OGX-427 in vitro and in vivo promotes AR ubiquitination and degradation and reduces tumor burden [93, 94]. Furthermore, OGX-427 reduces the metastatic potential of PC3 cells in vivo, validating a role for Hsp27 in cell migration and invasion [95]. As is the case for Hsp90 and Hsp70 inhibitors, it appears that inhibiting Hsp27 expression can sensitize cancer cells to chemotherapy [96, 97] and radiation [98], as well as to molecularly targeted therapies such as erlotinib [97]. Notably, OGX-427 synergizes with Hsp90 inhibitors to reduce prostate cancer growth [94]. In clinical trials, OGX-427 was well tolerated and led to PSA decline in a number of men with prostate cancer. Furthermore, those prostate cancer patients on an OGX-427 regimen were more likely to be progression-free after 12 weeks [92]. OGX-427 continues to be clinically evaluated, especially in men with advanced prostate cancer.

## 2.9 Small Hsps: Clusterin

Clusterin is another stress-inducible small Hsp that exerts cytoprotective effects. Like Hsp27, clusterin functions in an ATP-independent manner but demonstrates chaperone-like properties, regulating protein homeostasis and preventing the aggregation and precipitation of proteins [92]. By sequestering the pro-apoptotic factor Bax, clusterin also has anti-apoptotic activity, and its expression is associated with resistance to a number of therapeutic agents [99].

### *2.9.1 Antisense Oligonucleotides Targeting Clusterin*

As with OGX-427, clusterin can be targeted with an antisense oligonucleotide (OGX-011). In preclinical studies, OGX-011 promoted apoptosis of prostate cancer cells in vitro and inhibited tumor growth and metastatic potential in a prostate cancer xenograft model [100]. Due to clusterin's role in therapeutic resistance, it is not surprising that knockdown of clusterin expression with OGX-011 has synergistic activity when combined with radiation and chemotherapy [92]. Additionally, its combined activity with targeted agents such as enzalutamide, an AR antagonist, in prostate cancer is also greater than the activity of either agent alone [101]. Given clusterin's regulation by the Hsp90-induced HSR, co-administration with OGX-011 has been shown to markedly enhance the efficacy of the Hsp90 inhibitor, PF-04929113, to induce cell death and blunt AR (an Hsp90 client) expression and

**Table 2.1** A list of published chaperone inhibitors (not including Hsp90 inhibitors)

| Drug | Target | References |
|---|---|---|
| PU-H54, BnIm, KUNG29 | Grp94 | [39–42] |
| Selective ligands | FKBP51 | [46, 47] |
| Celastrol, gedunin | p23 | [48, 49] |
| VER-155008 | Hsc/Hsp70 | [51, 53, 54] |
| Apoptozole | Hsc/Hsp70 | [56–58] |
| MKT-077, YM-01, JG98 | Hsc/Hsp70 | [59–70, 72–74] |
| HS-72, YK5, MAL3-101 | Hsp70 | [75–77, 79] |
| Myricetin | DnaK | [78] |
| 115-7c | DnaK/DnaJ-DnaK complex | [80] |
| 116-9e | DnaJ-DnaK complex | [80] |
| PES, PES-Cl, PET-16 | Hsc/Hsp70 | [81–86] |
| ADD70 | Hsp70 | [87, 88] |
| Peptide aptamers | Hsp70 | [89] |
| Phenoxy-*N*-arylacetamides | DnaJ | [91] |
| OGX-427 | Hsp27 | [93–97] |
| OGX-011 | Clusterin | [100–103] |

transcriptional activity [100]. In patients undergoing radical prostatectomy, OGX-011 was well tolerated and decreased clusterin expression in tumor and lymph node tissue [102]. When combined with docetaxel and prednisone, OGX-011 provided a significant survival advantage in men with metastatic prostate cancer [103]. However, other studies have not confirmed these findings and OGX-011 continues to be clinically evaluated.

With the continued interest in developing small molecule and peptide-based chaperone inhibitors, an ever-expanding list of chaperone-targeted reagents has been developed. The impact of these diverse inhibitors on host immune function and their potential synergy or antagonism with immune-directed therapy remains an important subject for ongoing preclinical and clinical investigation (Table 2.1).

## References

1. Trepel J, Mollapour M, Giaccone G, Neckers L (2010) Targeting the dynamic HSP90 complex in cancer. Nat Rev Cancer 10:537–549
2. Graner MW (2016) HSP90 and immune modulation in cancer. Adv Cancer Res 129:191–224
3. Whitesell L, Mimnaugh EG, De Costa B, Myers CE, Neckers LM (1994) Inhibition of heat shock protein HSP90-pp60v-src heteroprotein complex formation by benzoquinone ansamycins: essential role for stress proteins in oncogenic transformation. Proc Natl Acad Sci U S A 91:8324–8328
4. Mimnaugh EG, Chavany C, Neckers L (1996) Polyubiquitination and proteasomal degradation of the p185c-erbB-2 receptor protein-tyrosine kinase induced by geldanamycin. J Biol Chem 271:22796–22801
5. Schulte TW, Akinaga S, Soga S, Sullivan W, Stensgard B, Toft D, Neckers LM (1998) Antibiotic radicicol binds to the N-terminal domain of Hsp90 and shares important biologic activities with geldanamycin. Cell Stress Chaperones 3:100–108

6. Workman P, Burrows F, Neckers L, Rosen N (2007) Drugging the cancer chaperone HSP90: combinatorial therapeutic exploitation of oncogene addiction and tumor stress. Ann N Y Acad Sci 1113:202–216
7. Grenert JP, Sullivan WP, Fadden P, Haystead TA, Clark J, Mimnaugh E, Krutzsch H, Ochel HJ, Schulte TW, Sausville E, Neckers LM, Toft DO (1997) The amino-terminal domain of heat shock protein 90 (hsp90) that binds geldanamycin is an ATP/ADP switch domain that regulates hsp90 conformation. J Biol Chem 272:23843–23850
8. Stebbins CE, Russo AA, Schneider C, Rosen N, Hartl FU, Pavletich NP (1997) Crystal structure of an Hsp90-geldanamycin complex: targeting of a protein chaperone by an antitumor agent. Cell 89:239–250
9. Roe SM, Prodromou C, O'Brien R, Ladbury JE, Piper PW, Pearl LH (1999) Structural basis for inhibition of the Hsp90 molecular chaperone by the antitumor antibiotics radicicol and geldanamycin. J Med Chem 42:260–266
10. Dutta R, Inouye M (2000) GHKL, an emergent ATPase/kinase superfamily. Trends Biochem Sci 25:24–28
11. Obermann WM, Sondermann H, Russo AA, Pavletich NP, Hartl FU (1998) In vivo function of Hsp90 is dependent on ATP binding and ATP hydrolysis. J Cell Biol 143:901–910
12. Panaretou B, Prodromou C, Roe SM, O'Brien R, Ladbury JE, Piper PW, Pearl LH (1998) ATP binding and hydrolysis are essential to the function of the Hsp90 molecular chaperone in vivo. EMBO J 17:4829–4836
13. Lang W, Caldwell GW, Li J, Leo GC, Jones WJ, Masucci JA (2007) Biotransformation of geldanamycin and 17-allylamino-17-demethoxygeldanamycin by human liver microsomes: reductive versus oxidative metabolism and implications. Drug Metab Dispos 35:21–29
14. Egorin MJ, Rosen DM, Wolff JH, Callery PS, Musser SM, Eiseman JL (1998) Metabolism of 17-(allylamino)-17-demethoxygeldanamycin (NSC 330507) by murine and human hepatic preparations. Cancer Res 58:2385–2396
15. Sidera K, Patsavoudi E (2014) HSP90 inhibitors: current development and potential in cancer therapy. Recent Pat Anticancer Drug Discov 9:1–20
16. Delmotte P, Delmotte-Plaque J (1953) A new antifungal substance of fungal origin. Nature 171:344
17. Khandelwal A, Crowley VM, Blagg BS (2016) Natural product inspired N-terminal Hsp90 inhibitors: from bench to bedside? Med Res Rev 36:92–118
18. Soga S, Neckers LM, Schulte TW, Shiotsu Y, Akasaka K, Narumi H, Agatsuma T, Ikuina Y, Murakata C, Tamaoki T, Akinaga S (1999) KF25706, a novel oxime derivative of radicicol, exhibits in vivo antitumor activity via selective depletion of Hsp90 binding signaling molecules. Cancer Res 59:2931–2938
19. Rowlands MG, Newbatt YM, Prodromou C, Pearl LH, Workman P, Aherne W (2004) High-throughput screening assay for inhibitors of heat-shock protein 90 ATPase activity. Anal Biochem 327:176–183
20. Chiosis G, Timaul MN, Lucas B, Munster PN, Zheng FF, Sepp-Lorenzino L, Rosen N (2001) A small molecule designed to bind to the adenine nucleotide pocket of Hsp90 causes Her2 degradation and the growth arrest and differentiation of breast cancer cells. Chem Biol 8:289–299
21. Chiosis G, Lucas B, Shtil A, Huezo H, Rosen N (2002) Development of a purine-scaffold novel class of Hsp90 binders that inhibit the proliferation of cancer cells and induce the degradation of Her2 tyrosine kinase. Bioorg Med Chem 10:3555–3564
22. Caldas-Lopes E, Cerchietti L, Ahn JH, Clement CC, Robles AI, Rodina A, Moulick K, Taldone T, Gozman A, Guo Y, Wu N, de Stanchina E, White J, Gross SS, Ma Y, Varticovski L, Melnick A, Chiosis G (2009) Hsp90 inhibitor PU-H71, a multimodal inhibitor of malignancy, induces complete responses in triple-negative breast cancer models. Proc Natl Acad Sci U S A 106:8368–8373
23. Trendowski M (2015) PU-H71: an improvement on nature's solutions to oncogenic Hsp90 addiction. Pharmacol Res 99:202–216
24. Immormino RM, Kang Y, Chiosis G, Gewirth DT (2006) Structural and quantum chemical studies of 8-aryl-sulfanyl adenine class Hsp90 inhibitors. J Med Chem 49:4953–4960

25. Patel HJ, Modi S, Chiosis G, Taldone T (2011) Advances in the discovery and development of heat-shock protein 90 inhibitors for cancer treatment. Expert Opin Drug Discov 6:559–587
26. Ohkubo S, Kodama Y, Muraoka H, Hitotsumachi H, Yoshimura C, Kitade M, Hashimoto A, Ito K, Gomori A, Takahashi K, Shibata Y, Kanoh A, Yonekura K (2015) TAS-116, a highly selective inhibitor of heat shock protein 90alpha and beta, demonstrates potent antitumor activity and minimal ocular toxicity in preclinical models. Mol Cancer Ther 14:14–22
27. Beebe K, Mollapour M, Scroggins B, Prodromou C, Xu W, Tokita M, Taldone T, Pullen L, Zierer BK, Lee MJ, Trepel J, Buchner J, Bolon D, Chiosis G, Neckers L (2013) Posttranslational modification and conformational state of heat shock protein 90 differentially affect binding of chemically diverse small molecule inhibitors. Oncotarget 4:1065–1074
28. Soti C, Vermes A, Haystead TA, Csermely P (2003) Comparative analysis of the ATP-binding sites of Hsp90 by nucleotide affinity cleavage: a distinct nucleotide specificity of the C-terminal ATP-binding site. Eur J Biochem 270:2421–2428
29. Marcu MG, Chadli A, Bouhouche I, Catelli M, Neckers LM (2000) The heat shock protein 90 antagonist novobiocin interacts with a previously unrecognized ATP-binding domain in the carboxyl terminus of the chaperone. J Biol Chem 275:37181–37186
30. Yu XM, Shen G, Neckers L, Blake H, Holzbeierlein J, Cronk B, Blagg BS (2005) Hsp90 inhibitors identified from a library of novobiocin analogues. J Am Chem Soc 127:12778–12779
31. Donnelly AC, Mays JR, Burlison JA, Nelson JT, Vielhauer G, Holzbeierlein J, Blagg BS (2008) The design, synthesis, and evaluation of coumarin ring derivatives of the novobiocin scaffold that exhibit antiproliferative activity. J Org Chem 73:8901–8920
32. Eskew JD, Sadikot T, Morales P, Duren A, Dunwiddie I, Swink M, Zhang X, Hembruff S, Donnelly A, Rajewski RA, Blagg BS, Manjarrez JR, Matts RL, Holzbeierlein JM, Vielhauer GA (2011) Development and characterization of a novel C-terminal inhibitor of Hsp90 in androgen dependent and independent prostate cancer cells. BMC Cancer 11:468
33. Yin Z, Henry EC, Gasiewicz TA (2009) (−)-Epigallocatechin-3-gallate is a novel Hsp90 inhibitor. Biochemistry 48:336–345
34. Khandelwal A, Hall JA, Blagg BS (2013) Synthesis and structure-activity relationships of EGCG analogues, a recently identified Hsp90 inhibitor. J Org Chem 78:7859–7884
35. Moses MA, Henry EC, Ricke WA, Gasiewicz TA (2015) The heat shock protein 90 inhibitor, (−)-epigallocatechin gallate, has anticancer activity in a novel human prostate cancer progression model. Cancer Prev Res (Phila) 8:249–257
36. Donnelly A, Blagg BS (2008) Novobiocin and additional inhibitors of the Hsp90 C-terminal nucleotide-binding pocket. Curr Med Chem 15:2702–2717
37. Gidalevitz T, Stevens F, Argon Y (2013) Orchestration of secretory protein folding by ER chaperones. Biochim Biophys Acta 1833:2410–2424
38. Sreedhar AS, Kalmar E, Csermely P, Shen YF (2004) Hsp90 isoforms: functions, expression and clinical importance. FEBS Lett 562:11–15
39. Patel PD, Yan P, Seidler PM, Patel HJ, Sun W, Yang C, Que NS, Taldone T, Finotti P, Stephani RA, Gewirth DT, Chiosis G (2013) Paralog-selective Hsp90 inhibitors define tumor-specific regulation of HER2. Nat Chem Biol 9:677–684
40. Duerfeldt AS, Peterson LB, Maynard JC, Ng CL, Eletto D, Ostrovsky O, Shinogle HE, Moore DS, Argon Y, Nicchitta CV, Blagg BS (2012) Development of a Grp94 inhibitor. J Am Chem Soc 134:9796–9804
41. Muth A, Crowley V, Khandelwal A, Mishra S, Zhao J, Hall J, Blagg BS (2014) Development of radamide analogs as Grp94 inhibitors. Bioorg Med Chem 22:4083–4098
42. Crowley VM, Khandelwal A, Mishra S, Stothert AR, Huard DJ, Zhao J, Muth A, Duerfeldt AS, Kizziah JL, Lieberman RL, Dickey CA, Blagg BS (2016) Development of glucose regulated protein 94-selective inhibitors based on the BnIm and radamide scaffold. J Med Chem 59:3471–3488
43. Smith JR, Clarke PA, de Billy E, Workman P (2009) Silencing the cochaperone CDC37 destabilizes kinase clients and sensitizes cancer cells to HSP90 inhibitors. Oncogene 28:157–169
44. Polier S, Samant RS, Clarke PA, Workman P, Prodromou C, Pearl LH (2013) ATP-competitive inhibitors block protein kinase recruitment to the Hsp90-Cdc37 system. Nat Chem Biol 9:307–312

45. Wang W, Liu Y, Zhao Z, Xie C, Xu Y, Hu Y, Quan H, Lou L (2016) Y-632 inhibits Hsp90 function through disrupting the interaction of Hsp90-Hop and exerts antitumor activity in vitro and in vivo. Cancer Sci 107(6):782–790
46. Baughman G, Wiederrecht GJ, Campbell NF, Martin MM, Bourgeois S (1995) FKBP51, a novel T-cell-specific immunophilin capable of calcineurin inhibition. Mol Cell Biol 15:4395–4402
47. Gaali S, Kirschner A, Cuboni S, Hartmann J, Kozany C, Balsevich G, Namendorf C, Fernandez-Vizarra P, Sippel C, Zannas AS, Draenert R, Binder EB, Almeida OF, Ruhter G, Uhr M, Schmidt MV, Touma C, Bracher A, Hausch F (2015) Selective inhibitors of the FK506-binding protein 51 by induced fit. Nat Chem Biol 11:33–37
48. Chadli A, Felts SJ, Wang Q, Sullivan WP, Botuyan MV, Fauq A, Ramirez-Alvarado M, Mer G (2010) Celastrol inhibits Hsp90 chaperoning of steroid receptors by inducing fibrillization of the Co-chaperone p23. J Biol Chem 285:4224–4231
49. Patwardhan CA, Fauq A, Peterson LB, Miller C, Blagg BS, Chadli A (2013) Gedunin inactivates the co-chaperone p23 protein causing cancer cell death by apoptosis. J Biol Chem 288:7313–7325
50. Sondermann H, Scheufler C, Schneider C, Hohfeld J, Hartl FU, Moarefi I (2001) Structure of a Bag/Hsc70 complex: convergent functional evolution of Hsp70 nucleotide exchange factors. Science 291:1553–1557
51. Williamson DS, Borgognoni J, Clay A, Daniels Z, Dokurno P, Drysdale MJ, Foloppe N, Francis GL, Graham CJ, Howes R, Macias AT, Murray JB, Parsons R, Shaw T, Surgenor AE, Terry L, Wang Y, Wood M, Massey AJ (2009) Novel adenosine-derived inhibitors of 70 kDa heat shock protein, discovered through structure-based design. J Med Chem 52:1510–1513
52. Assimon VA, Gillies AT, Rauch JN, Gestwicki JE (2013) Hsp70 protein complexes as drug targets. Curr Pharm Des 19:404–417
53. Massey AJ, Williamson DS, Browne H, Murray JB, Dokurno P, Shaw T, Macias AT, Daniels Z, Geoffroy S, Dopson M, Lavan P, Matassova N, Francis GL, Graham CJ, Parsons R, Wang Y, Padfield A, Comer M, Drysdale MJ, Wood M (2010) A novel, small molecule inhibitor of Hsc70/Hsp70 potentiates Hsp90 inhibitor induced apoptosis in HCT116 colon carcinoma cells. Cancer Chemother Pharmacol 66:535–545
54. Boroughs LK, Antonyak MA, Johnson JL, Cerione RA (2011) A unique role for heat shock protein 70 and its binding partner tissue transglutaminase in cancer cell migration. J Biol Chem 286:37094–37107
55. Santos F, Nequiz M, Hernandez-Cuevas NA, Hernandez K, Pineda E, Encalada R, Guillen N, Luis-Garcia E, Saralegui A, Saavedra E, Perez-Tamayo R, Olivos-Garcia A (2015) Maintenance of intracellular hypoxia and adequate heat shock response are essential requirements for pathogenicity and virulence of Entamoeba histolytica. Cell Microbiol 17:1037–1051
56. Williams DR, Ko SK, Park S, Lee MR, Shin I (2008) An apoptosis-inducing small molecule that binds to heat shock protein 70. Angew Chem Int Ed Engl 47:7466–7469
57. Evans LE, Cheeseman MD, Yahya N, Jones K (2015) Investigating apoptozole as a chemical probe for HSP70 inhibition. PLoS One 10:e0140006
58. Ko SK, Kim J, Na DC, Park S, Park SH, Hyun JY, Baek KH, Kim ND, Kim NK, Park YN, Song K, Shin I (2015) A small molecule inhibitor of ATPase activity of HSP70 induces apoptosis and has antitumor activities. Chem Biol 22:391–403
59. Koya K, Li Y, Wang H, Ukai T, Tatsuta N, Kawakami M, Shishido, Chen LB (1996) MKT-077, a novel rhodacyanine dye in clinical trials, exhibits anticarcinoma activity in preclinical studies based on selective mitochondrial accumulation. Cancer Res 56:538–543
60. Chiba Y, Kubota T, Watanabe M, Otani Y, Teramoto T, Matsumoto Y, Koya K, Kitajima M (1998) Selective antitumor activity of MKT-077, a delocalized lipophilic cation, on normal cells and cancer cells in vitro. J Surg Oncol 69:105–110
61. Tikoo A, Shakri R, Connolly L, Hirokawa Y, Shishido T, Bowers B, Ye LH, Kohama K, Simpson RJ, Maruta H (2000) Treatment of ras-induced cancers by the F-actin-bundling drug MKT-077. Cancer J 6:162–168

62. Wadhwa R, Sugihara T, Yoshida A, Nomura H, Reddel RR, Simpson R, Maruta H, Kaul SC (2000) Selective toxicity of MKT-077 to cancer cells is mediated by its binding to the hsp70 family protein mot-2 and reactivation of p53 function. Cancer Res 60:6818–6821
63. Rousaki A, Miyata Y, Jinwal UK, Dickey CA, Gestwicki JE, Zuiderweg ER (2011) Allosteric drugs: the interaction of antitumor compound MKT-077 with human Hsp70 chaperones. J Mol Biol 411:614–632
64. Chiba Y, Kubota T, Watanabe M, Matsuzaki SW, Otani Y, Teramoto T, Matsumoto Y, Koya K, Kitajima M (1998) MKT-077, localized lipophilic cation: antitumor activity against human tumor xenografts serially transplanted into nude mice. Anticancer Res 18:1047–1052
65. Tatsuta N, Suzuki N, Mochizuki T, Koya K, Kawakami M, Shishido T, Motoji N, Kuroiwa H, Shigematsu A, Chen LB (1999) Pharmacokinetic analysis and antitumor efficacy of MKT-077, a novel antitumor agent. Cancer Chemother Pharmacol 43:295–301
66. Wadhwa R, Colgin L, Yaguchi T, Taira K, Reddel RR, Kaul SC (2002) Rhodacyanine dye MKT-077 inhibits in vitro telomerase assay but has no detectable effects on telomerase activity in vivo. Cancer Res 62:4434–4438
67. Koren J 3rd, Miyata Y, Kiray J, O'Leary JC 3rd, Nguyen L, Guo J, Blair LJ, Li X, Jinwal UK, Cheng JQ, Gestwicki JE, Dickey CA (2012) Rhodacyanine derivative selectively targets cancer cells and overcomes tamoxifen resistance. PLoS One 7:e35566
68. Abisambra J, Jinwal UK, Miyata Y, Rogers J, Blair L, Li X, Seguin SP, Wang L, Jin Y, Bacon J, Brady S, Cockman M, Guidi C, Zhang J, Koren J, Young ZT, Atkins CA, Zhang B, Lawson LY, Weeber EJ, Brodsky JL, Gestwicki JE, Dickey CA (2013) Allosteric heat shock protein 70 inhibitors rapidly rescue synaptic plasticity deficits by reducing aberrant tau. Biol Psychiatry 74:367–374
69. Colvin TA, Gabai VL, Gong J, Calderwood SK, Li H, Gummuluru S, Matchuk ON, Smirnova SG, Orlova NV, Zamulaeva IA, Garcia-Marcos M, Li X, Young ZT, Rauch JN, Gestwicki JE, Takayama S, Sherman MY (2014) Hsp70-Bag3 interactions regulate cancer-related signaling networks. Cancer Res 74:4731–4740
70. Wang AM, Miyata Y, Klinedinst S, Peng HM, Chua JP, Komiyama T, Li X, Morishima Y, Merry DE, Pratt WB, Osawa Y, Collins CA, Gestwicki JE, Lieberman AP (2013) Activation of Hsp70 reduces neurotoxicity by promoting polyglutamine protein degradation. Nat Chem Biol 9:112–118
71. Pratt WB, Gestwicki JE, Osawa Y, Lieberman AP (2015) Targeting Hsp90/Hsp70-based protein quality control for treatment of adult onset neurodegenerative diseases. Annu Rev Pharmacol Toxicol 55:353–371
72. Li X, Srinivasan SR, Connarn J, Ahmad A, Young ZT, Kabza AM, Zuiderweg ER, Sun D, Gestwicki JE (2013) Analogs of the allosteric heat shock protein 70 (Hsp70) inhibitor, MKT-077, as anti-cancer agents. ACS Med Chem Lett 4. https://doi.org/10.1021/ml400204n
73. Li X, Colvin T, Rauch JN, Acosta-Alvear D, Kampmann M, Dunyak B, Hann B, Aftab BT, Murnane M, Cho M, Walter P, Weissman JS, Sherman MY, Gestwicki JE (2015) Validation of the Hsp70-Bag3 protein-protein interaction as a potential therapeutic target in cancer. Mol Cancer Ther 14:642–648
74. Taguwa S, Maringer K, Li X, Bernal-Rubio D, Rauch JN, Gestwicki JE, Andino R, Fernandez-Sesma A, Frydman J (2015) Defining Hsp70 subnetworks in dengue virus replication reveals key vulnerability in flavivirus infection. Cell 163:1108–1123
75. Howe MK, Bodoor K, Carlson DA, Hughes PF, Alwarawrah Y, Loiselle DR, Jaeger AM, Darr DB, Jordan JL, Hunter LM, Molzberger ET, Gobillot TA, Thiele DJ, Brodsky JL, Spector NL, Haystead TA (2014) Identification of an allosteric small-molecule inhibitor selective for the inducible form of heat shock protein 70. Chem Biol 21:1648–1659
76. Howe MK, Speer BL, Hughes PF, Loiselle DR, Vasudevan S, Haystead TA (2016) An inducible heat shock protein 70 small molecule inhibitor demonstrates anti-dengue virus activity, validating Hsp70 as a host antiviral target. Antivir Res 130:81–92
77. Rodina A, Patel PD, Kang Y, Patel Y, Baaklini I, Wong MJ, Taldone T, Yan P, Yang C, Maharaj R, Gozman A, Patel MR, Patel HJ, Chirico W, Erdjument-Bromage H, Talele TT, Young JC, Chiosis G (2013) Identification of an allosteric pocket on human hsp70 reveals a mode of inhibition of this therapeutically important protein. Chem Biol 20:1469–1480

78. Chang L, Miyata Y, Ung PM, Bertelsen EB, McQuade TJ, Carlson HA, Zuiderweg ER, Gestwicki JE (2011) Chemical screens against a reconstituted multiprotein complex: myricetin blocks DnaJ regulation of DnaK through an allosteric mechanism. Chem Biol 18:210–221
79. Fewell SW, Smith CM, Lyon MA, Dumitrescu TP, Wipf P, Day BW, Brodsky JL (2004) Small molecule modulators of endogenous and co-chaperone-stimulated Hsp70 ATPase activity. J Biol Chem 279:51131–51140
80. Wisen S, Bertelsen EB, Thompson AD, Patury S, Ung P, Chang L, Evans CG, Walter GM, Wipf P, Carlson HA, Brodsky JL, Zuiderweg ER, Gestwicki JE (2010) Binding of a small molecule at a protein-protein interface regulates the chaperone activity of hsp70-hsp40. ACS Chem Biol 5:611–622
81. Strom E, Sathe S, Komarov PG, Chernova OB, Pavlovska I, Shyshynova I, Bosykh DA, Burdelya LG, Macklis RM, Skaliter R, Komarova EA, Gudkov AV (2006) Small-molecule inhibitor of p53 binding to mitochondria protects mice from gamma radiation. Nat Chem Biol 2:474–479
82. Leu JI, Pimkina J, Frank A, Murphy ME, George DL (2009) A small molecule inhibitor of inducible heat shock protein 70. Mol Cell 36:15–27
83. Balaburski GM, Leu JI, Beeharry N, Hayik S, Andrake MD, Zhang G, Herlyn M, Villanueva J, Dunbrack RL Jr, Yen T, George DL, Murphy ME (2013) A modified HSP70 inhibitor shows broad activity as an anticancer agent. Mol Cancer Res 11:219–229
84. Leu JI, Zhang P, Murphy ME, Marmorstein R, George DL (2014) Structural basis for the inhibition of HSP70 and DnaK chaperones by small-molecule targeting of a C-terminal allosteric pocket. ACS Chem Biol 9:2508–2516
85. Zhang P, Leu JI, Murphy ME, George DL, Marmorstein R (2014) Crystal structure of the stress-inducible human heat shock protein 70 substrate-binding domain in complex with peptide substrate. PLoS One 9:e103518
86. Leu JI, Pimkina J, Pandey P, Murphy ME, George DL (2011) HSP70 inhibition by the small-molecule 2-phenylethynesulfonamide impairs protein clearance pathways in tumor cells. Mol Cancer Res 9:936–947
87. Schmitt E, Parcellier A, Gurbuxani S, Cande C, Hammann A, Morales MC, Hunt CR, Dix DJ, Kroemer RT, Giordanetto F, Jaattela M, Penninger JM, Pance A, Kroemer G, Garrido C (2003) Chemosensitization by a non-apoptogenic heat shock protein 70-binding apoptosis-inducing factor mutant. Cancer Res 63:8233–8240
88. Schmitt E, Maingret L, Puig PE, Rerole AL, Ghiringhelli F, Hammann A, Solary E, Kroemer G, Garrido C (2006) Heat shock protein 70 neutralization exerts potent antitumor effects in animal models of colon cancer and melanoma. Cancer Res 66:4191–4197
89. Rerole AL, Gobbo J, De Thonel A, Schmitt E, Pais de Barros JP, Hammann A, Lanneau D, Fourmaux E, Demidov ON, Micheau O, Lagrost L, Colas P, Kroemer G, Garrido C (2011) Peptides and aptamers targeting HSP70: a novel approach for anticancer chemotherapy. Cancer Res 71:484–495
90. Dekker SL, Kampinga HH, Bergink S (2015) DNAJs: more than substrate delivery to HSPA. Front Mol Biosci 2:35
91. Cassel JA, Ilyin S, McDonnell ME, Reitz AB (2012) Novel inhibitors of heat shock protein Hsp70-mediated luciferase refolding that bind to DnaJ. Bioorg Med Chem 20:3609–3614
92. Azad AA, Zoubeidi A, Gleave ME, Chi KN (2015) Targeting heat shock proteins in metastatic castration-resistant prostate cancer. Nat Rev Urol 12:26–36
93. Zoubeidi A, Zardan A, Beraldi E, Fazli L, Sowery R, Rennie P, Nelson C, Gleave M (2007) Cooperative interactions between androgen receptor (AR) and heat-shock protein 27 facilitate AR transcriptional activity. Cancer Res 67:10455–10465
94. Lamoureux F, Thomas C, Yin MJ, Fazli L, Zoubeidi A, Gleave ME (2014) Suppression of heat shock protein 27 using OGX-427 induces endoplasmic reticulum stress and potentiates heat shock protein 90 inhibitors to delay castrate-resistant prostate cancer. Eur Urol 66:145–155
95. Shiota M, Bishop JL, Nip KM, Zardan A, Takeuchi A, Cordonnier T, Beraldi E, Bazov J, Fazli L, Chi K, Gleave M, Zoubeidi A (2013) Hsp27 regulates epithelial mesenchymal transition, metastasis, and circulating tumor cells in prostate cancer. Cancer Res 73:3109–3119

96. Baylot V, Andrieu C, Katsogiannou M, Taieb D, Garcia S, Giusiano S, Acunzo J, Iovanna J, Gleave M, Garrido C, Rocchi P (2011) OGX-427 inhibits tumor progression and enhances gemcitabine chemotherapy in pancreatic cancer. Cell Death Dis 2:e221
97. Lelj-Garolla B, Kumano M, Beraldi E, Nappi L, Rocchi P, Ionescu DN, Fazli L, Zoubeidi A, Gleave ME (2015) Hsp27 inhibition with OGX-427 sensitizes non-small cell lung cancer cells to erlotinib and chemotherapy. Mol Cancer Ther 14:1107–1116
98. Hadchity E, Aloy MT, Paulin C, Armandy E, Watkin E, Rousson R, Gleave M, Chapet O, Rodriguez-Lafrasse C (2009) Heat shock protein 27 as a new therapeutic target for radiation sensitization of head and neck squamous cell carcinoma. Mol Ther 17:1387–1394
99. Zhang H, Kim JK, Edwards CA, Xu Z, Taichman R, Wang CY (2005) Clusterin inhibits apoptosis by interacting with activated Bax. Nat Cell Biol 7:909–915
100. Lamoureux F, Thomas C, Yin MJ, Kuruma H, Beraldi E, Fazli L, Zoubeidi A, Gleave ME (2011) Clusterin inhibition using OGX-011 synergistically enhances Hsp90 inhibitor activity by suppressing the heat shock response in castrate-resistant prostate cancer. Cancer Res 71:5838–5849
101. Matsumoto H, Yamamoto Y, Shiota M, Kuruma H, Beraldi E, Matsuyama H, Zoubeidi A, Gleave M (2013) Cotargeting androgen receptor and clusterin delays castrate-resistant prostate cancer progression by inhibiting adaptive stress response and AR stability. Cancer Res 73:5206–5217
102. Chi KN, Eisenhauer E, Fazli L, Jones EC, Goldenberg SL, Powers J, Tu D, Gleave ME (2005) A phase I pharmacokinetic and pharmacodynamic study of OGX-011, a 2′-methoxyethyl antisense oligonucleotide to clusterin, in patients with localized prostate cancer. J Natl Cancer Inst 97:1287–1296
103. Chi KN, Hotte SJ, Yu EY, Tu D, Eigl BJ, Tannock I, Saad F, North S, Powers J, Gleave ME, Eisenhauer EA (2010) Randomized phase II study of docetaxel and prednisone with or without OGX-011 in patients with metastatic castration-resistant prostate cancer. J Clin Oncol 28:4247–4254

# Part II
# Exposure of HSPs to Immune Cells

# Chapter 3
# Extracellular Heat Shock Proteins as Stress Communication Signals

**Antonio De Maio**

**Abstract** Intercellular communication is a fundamental process necessary to maintain homeostasis and to mount an orchestrated response to stress. Although heat shock proteins (HSP) play a critical role by participating in the repair of damaged products as a result of the stress in the intracellular milieu, it is now evident that they play an alternative role when they escape from the cells and are placed in circulation, participating in a systemic stress response. Extracellular HSP appear as signaling molecules involved in intercellular communication during stress conditions. They have been found to modulate the function of many target cells. Moreover, extracellular HSP have been detected in several biological fluids, particularly from patients suffering from a large number of maladies. Extracellular HSP are released by many cell types and by several mechanisms, including passive dissemination after necrosis and active export by a nonclassical secretory pathway. Among several potential mechanisms for the export of HSP, their release associated with extracellular vesicles has gained increasing support. The appearance of extracellular vesicles containing HSP emerges as a new form of cellular communication during stress conditions directed at avoiding the propagation of the insult.

**Keywords** Heat shock proteins · Cellular communication · Extracellular vesicles · Stress · Signaling

A. De Maio (✉)
Department of Surgery, School of Medicine, University of California San Diego, La Jolla, CA, USA

Department of Neuroscience, School of Medicine, University of California San Diego, La Jolla, CA, USA

Center for Investigations of Health and Education Disparities, University of California San Diego, La Jolla, CA, USA
e-mail: ademaio@ucsd.edu



R. J. Binder, P. K. Srivastava (eds.), *Heat Shock Proteins in the Immune System*,
https://doi.org/10.1007/978-3-319-69042-1_3

## 3.1 Introduction

Cellular communication is a major physiological event that is crucial to maintain homeostasis. In this regard, unicellular organisms use chemical gradients to synchronize their metabolism and growth. Plants release volatile compounds as signals to coordinate development, attract pollinizing insects, and repel predators. Multicellular organisms send signals to their counterpart cells to regulate metabolism, growth, and stress response. Intercellular communication is particularly important in tissue homeostasis, in which a synchronized propagation of signals among cells is required to adapt to changes in nutrients and other environmental factors. For example, hepatocytes within the hepatic acinus are interconnected by various mechanisms of communication to modulate their response to changes in the delivery of nutrients, such as glucose and oxygen. Cellular communication is also critical for an efficient response of the immune system. Thus, the communication between T, B, and antigen-presenting cells is necessary to orchestrate the adaptive immune response. Similarly, macrophages, dendritic cells, and neutrophils secrete cytokines and chemokines in response to infection and injury to promote an initial response to the insult. Therefore, a coordinated intercellular communication is vital to preserve normal physiological conditions and mount a sound response to stress.

## 3.2 Types of Cellular Communication

Cells communicate by a variety of mechanisms. The most common is via soluble molecules that are placed in the extracellular environment or in circulation directed at interacting with adjacent or distant cells. A typical example of this type of communication is when hormones and cytokines are released by a certain type of cell and captured by another via specific receptors. The ligand-receptor interaction triggers a signal transduction pathway within the plasma membrane or within internal compartments directed at activating the right response to the external stimuli. In other cases, cells communicate via surface contact molecules, such as adhesion proteins, resulting in cellular synapses [1]. A great example is the immune synapse between antigen presenting cells and lymphocytes [2]. Cells in close proximity can also interact via exchanging surface molecules by the direct transfer of plasma membrane portions, which is known as trogocytosis, by membrane tethers, or by nanotubes [3]. A very important form of cellular communication is via the transfer of low-molecular-weight metabolites via gap junctions. Gap junctions are larger channels or pores formed by similar proteins within the plasma membrane of adjacent cells that allow the passage of ions (e.g., calcium), nucleotides (e.g., ATP), and other small molecules in a regulated process, creating a network of signals across the multicellular environment [4].

An alternative mechanism for cellular communication is mediated by the release of membrane vesicles into the extracellular medium. These extracellular vesicles

(ECV) contain surface molecules, lipids, and cargo (e.g., proteins, nucleic acids, carbohydrates, and ions). The critical aspect of ECV is that they contain a large number of molecules within a very small volume [5–7]. These ECV are captured by target cells delivering the cargo, such as signaling proteins or microRNAs, which can modulate the function of the receiving cell. More importantly, the target cell senses a multiplicity of different molecules simultaneously, which is likely to result in a synergy of information. In other words, different components within ECV could concurrently activate various cellular pathways. Moreover, the concentration of a ligand within a small vesicle (e.g., 100 nm in diameter) is theoretically calculated in the millimolar range, which is much larger than the circulating concentration of any hormone. ECV could also travel long distances, delivering information to very distant cells. The fact that ECV are formed by membrane-encapsulated macromolecules assures the protection of the cargo from external environmental factors, such as circulating proteases and RNAses. The final stage for communication via ECV requires the recognition by the target cell that it could be mediated by various mechanisms. For example, ECV may contain surface molecules that are recognized by specific receptors on target cells acting as "zip codes." In addition, ECV could be taken by cells in a non-receptor-mediated process, such as macropinocytosis, or they could fuse with the plasma membrane delivering the cargo into the cytosol.

## 3.3 Extracellular HSP as Communication Signals

Heat shock proteins (HSP) were first discovered as part of the cellular response to elevated temperatures, initiated by the discovery of the heat shock response by Ritossa [8], followed by the identification of HSP by Tissières et al. [9]. Subsequent studies showed that HSP correspond to a large family of proteins expressed after a variety of stress conditions [10, 11]. Various homologs to the stress-inducible HSP were identified afterward participating in normal basic cellular processes, including folding of newly synthesized polypeptides, translocation of polypeptides across subcellular compartments, assembly of macromolecule complexes, stabilization of receptors, and signal transduction [11, 12]. The capability of folding denatured proteins or stabilizing protein complexes gave rise to their denomination as molecular chaperones [13]. Various HSP belong to particular families that are classified according to their molecular weight, for example the Hsp70 family, which has a molecular weight of 70 kDa, is composed of four members: Hsp70 (the stress inducible form), Hsc70 (the constitutive cytosolic form), Mit70 and Grp78 (both constitutive forms located in the mitochondria and endoplasmic reticulum, respectively). Recently, a new nomenclature for HSP has been proposed [14], displayed in Table 3.1.

Although the most recognized function of HSP is as molecular chaperones in the cytosol and other subcellular compartments, they have been found outside cells. The first observations regarding the presence of HSP in the extracellular environment was made on Hsp70 by studies of Tytell et al. [15] and Hightower and Guidon

**Table 3.1** Classification of HSP

| Family name | Common name | New name |
|---|---|---|
| HSP 100 | HSP105 | HSPH1 |
| | HSP110 | HSPH2 |
| | Grp170 | HSPH4 |
| HSP90 | HSP90α | HSPC2 |
| | HSP90β | HSPC3 |
| | Grp94 | HSPC4 |
| HSP70 | HSP70(HSP72) | HSPA1 |
| | HSC70(HSP73) | HSPAB |
| | Grp78(BIP) | HSPA5 |
| | Utp70 (Grp75) | HSPA9 |
| HSP40 | HSP40 (Dnaj) | DNAJB1 |
| Small HSP | αA-Crystallin | HSPB4 |
| | αB-Crystallin | HSPB5 |
| | HSP25 | HSPB1 |
| | HSP27 | HSPB2 |
| | HSP20 | HSPB6 |
| | HSP22 | HSPB8 |
| Chaperonins | GloEL (HSP60) | HSPD1 |
| | GloES | HSPE1 |

[16]. These pioneer findings were followed by more recent observations documenting the presence of Hsp70 in the extracellular medium in a variety of conditions (reviewed by De Maio) [6]. Today, practically all members of the HSP family have been detected outside cells. Thus, Hsp90α (HSPC3) was identified as a secreted oxidative stress-induced factor by vascular smooth cells [17]. Hsp90β (HSPC4) was reported released by osteosarcoma cells [18]. Grp75 (HSPA9) or mortalin, which is a mitochondrial chaperone protein, has been shown to be released after complement treatment of cells [19]. Grp78 (HSPA5) and Grp94 (HSPC4), which are endoplasmic reticulum (ER) residents, have been found in the extracellular space [20–22]. HSP60 (HSPD1) has been detected in circulation of patients suffering from various conditions [23]. Hsp25/27 (HspB1) was observed in the extracellular environment of astrocytes [24]. Finally, a large member of the HSP family, Grp170 (HSPH4), has also been detected outside cells [25].

Extracellular HSP are secreted by a variety of cell types and captured by others. The function of extracellular HSP has not been associated to their chaperone activity, which is not surprising since it requires cochaperones and nucleotides for the function. On the contrary, extracellular HSP act as signaling molecules involved in the communication between cells, inducing an array of activities. For example, Hsp70 (HSPA1) secreted by parenchymal cells has been shown to induce a robust activation of macrophages [26–28], dendritic cells [29], and natural killer cells [30, 31]. Extracellular Hsp70 has also been shown to modulate the response of monocytes to endotoxin [32, 33], activate chemotaxis [34], and phagocytosis [35–37].

They also could modulate antigen presentation [36]. *Mycobacterium tuberculosis*-derived DnaK has been shown to polarize macrophages to M2-like phenotype [38] and induce anti-inflammatory response [39]. Recently, Hsp70, Hsp90 (HSPC), and Hsp40 (DNAJB1) have been proposed to promote protein homeostasis in distant cells [40]. Moreover, extracellular Hsp70 has been associated with both immunostimulatory and immunosuppressive activities [41]. Extracellular Hsp90 has been shown to transport antigens from the outside to the cytosol, resulting in cross-presentation [42]. Small HSP are also secreted by cells and modulate the immune system [43]. Hsp90α was detected outside cells participating in wound reepithelialization and healing [37, 44, 45]. Extracellular Hsp70 has been shown to affect cardiomyocyte contractile dysfunction [46], and increase tumor growth and resistance to apoptosis [46]. Exogenous Hsp70 appeared to disrupt gap junction communication between human microvascular endothelial cells [47].

Extracellular HSP may be recognized by a variety of cell surface targets [48]. The list of potential receptors for extracellular HSP is large, including CD91 [49, 50], CD40 [13, 51], Scavenger receptor A [52], Lox 1 [53], mannose receptor [54], and even the β-subunit of ATP synthase [55]. Recently, Hsp70 was shown to bind to Siglec-5 and Siglec-14, which are Ig-superfamily lectins on mammalian leukocytes that recognize sialic acid-bearing glycans [56]. Some lipids have also been proposed as targets for HSP, such as sphingolipids [57, 58], phosphatidylserine [59, 60] and phospholipid bis(monoacylglycero)phosphate [61]. In general, it appears that HSP have a large appetite for molecules, raising the possibility that a single receptor model may not be correct.

## 3.4 Extracellular HSP in Pathological Conditions

Extracellular HSP has been associated with several clinical conditions, following their detection in various biological fluids (Table 3.2). In addition, antibodies against HSP have been found in the serum of a variety of patients [23, 62]. For example, circulating levels of Hsp70 and Hsp60 or their antibodies have been proposed as a risk factor for coronary heart disease [63–65]. Similarly, Hsp60 has been detected in circulation of individuals suffering from cardiovascular diseases [66, 67]. Extracellular Hsp25 has been shown to reduce cardiotoxicity induced by doxorubicin [68]. Hsp70 has been reported to be present in the serum of patients with chronic hepatitis, liver cirrhosis, and hepatocellular carcinoma [69]. Hsp27 has been detected in the serum of patients with chronic pancreatitis and pancreatic carcinoma [70, 71]. Hsp60 has also been observed in the saliva and serum of patients with type 2 diabetes mellitus [72] and Hsp70 in patients presenting diabetic ketoacidosis [73].

Extracellular HSP have been associated with infection and other pathologies. Thus, extracellular Hsp70 has been identified following acute infection in humans [86]. Hsp70 and Hsp60 were found in wound fluids at the site of soft tissue injury [83]. Moreover, the presence of Hsp70 in circulation has also been linked with improved survival of critically ill patients [80–82]. In other studies, Hsp70 was

**Table 3.2** Heat shock proteins in clinical conditions

| Pathology | HSP | Reference |
|---|---|---|
| Heart disease | Hsp70, Hsp60 | [66]<br>[63]<br>[74]<br>[75]<br>[64]<br>[65]<br>[46] |
| Cancer | Hsp70, Hsc70 | [76]<br>[77]<br>[78]<br>[69] |
| Liver cirrhosis | Hsp70 | [69] |
| Hepatitis | | [69] |
| Pancreatitis | Hsp27 | [71]<br>[70] |
| Diabetes | Hsp60, Hsp70 | [79]<br>[73]<br>[72] |
| Trauma | Hsp70, Hsp60 | [80]<br>[81]<br>[82]<br>[83] |
| Ischemia–reperfusion injury | Hsp70 | [75]<br>[84]<br>[85] |
| Infection | Hsp70 | [86]<br>[87] |
| Preeclampsia | Hsp70 | [88]<br>[89] |

found to be released from human fetal membranes after exposure to *E. coli* [87]. Hsp70 has also been detected in normal and pathological pregnancies, including preeclampsia [88, 89]. Moreover, Hsp70 has been observed in amniotic fluid [90]. Circulating Hsp70 has been detected after extenuating exercise [91–93]. Finally, extracellular Hsp70 was present in the blood of experimental rodent models of diabetes [94], sepsis [95], and ischemia–reperfusion injury [85].

The central nervous system has also been a target for extracellular HSP activity. For example, extracellular Hsp70 has been observed after brain and spinal cord ischemia [84]. Several small HSP, including Hsp20 and Hsp22, have been detected in closed proximity of amyloid β deposits in the brains of Alzheimer disease patients [96]. Moreover, they were found to block amyloid β aggregation in vitro [96, 97]. Similar observations regarding inhibition of amyloid β aggregation have been made for Hsp70 and Hsp90 [98] and Hsp40 [99]. Moreover, HspB1 (Hsp25/27) was reported released by astrocytes in response to amyloid β exposure [24].

Extracellular Hsp70 has been shown to protect Schwann cells [100] and neurons [101]. Both αA-crystallin (HSPB4) and αB-crystallin (HSPB5) have been reported to protect astrocytes from various toxic agents [102].

## 3.5 Mechanisms of HSP Export

Extracellular HSP are released from at least two different sources. HSP are disseminated by a passive process secondary to cell lysis after necrosis [29, 85] or exported by an active mechanism independent of cell death, which could not be blocked by typical inhibitors of the ER-Golgi pathway, such as brefeldin A [16, 79]. The only exceptions are ER-resident HSP, Grp78 and Grp94, which are already in place within the classical secretory pathway. In contrast, the majority of HSP lack the consensus signal required for secretion via the ER-Golgi pathway. Therefore, they are likely exported by an alternative mechanism that has been named the nonclassical or unconventional secretory pathway. Many proteins besides HSP use this pathway for export, including interleukin-1β, high-mobility group box 1, galectin 1 and 3, and fibroblast growth factor 1 [103].

The major argument against the active export of HSP is that these proteins are localized in the cytosol. In order to reach the extracellular environment, they need to cross the plasma membrane. Thermodynamically, the passing of a protein across a lipid membrane results in a less favorable change of free energy [104]. Therefore, it should not be a spontaneous process. In spite of the prior assumption, there is substantial evidence that Hsp70s can spontaneously get inserted into lipid bilayers. Indeed, our pioneering work showed that Hsc70 got inserted into artificial lipid bilayers opening ion conductance pathways [105]. This observation has been confirmed by others [106] and extended to Hsp70 [28]. Additional studies showed that both Hsp70 and Hsc70 interact with liposomes resulting in their aggregation in a process that was concentration dependent and requiring the presence of nucleotides [4]. Moreover, Hsp70 have been demonstrated to get spontaneously and selectively inserted into phosphatidylserine (PS) liposomes, forming high molecular mass oligomers [107]. The interaction of Hsp70 with PS liposomes has been confirmed by others [108]. Similarly bacterial Hsp70 (DnaK) also gets inserted into liposomes, but the translocation was not lipid specific and only forms dimeric forms within the membrane [109]. These observations suggest that HSP, at least Hsp70, could move spontaneously from the cytosol into the plasma membrane. Indeed, Hsp70 has been extensively reported to be present in the plasma membrane of transformed cells [110, 111]. The presence of Hsp70 on the plasma membrane was resistant to acid washes indicating that it was actually inserted into the lipid bilayer [28, 57]. Therefore, the question that emerges is whether Hsp70 could also spontaneously come out from the lipid membrane outside the cells. Although this option has not been demonstrated experimentally, it may be an interesting possibility to explain the extracellular release of this protein. Other mechanisms that have been proposed for the active secretion of HSP include the lysosome–endosome pathway, in which the protein translocates into the lysosome lumen via ATP-binding cassette (ABC) transport-like system and is further released outside cells via the endocytic process [112]. This pathway has also been proposed for the secretion of IL-1β, which also moves from the cytosol to outside the cell without passing through the ER-Golgi pathway [113]. Other studies have suggested the release of Hsp70 via secretory-like granules [114].

A well-accepted mechanism for the export of HSP is their association with ECV [6]. These vesicles are derived from the plasma membrane by various processes. Protuberances or blebs can be formed in the outer side of the plasma membrane by a process of ectocytosis, resulting in the release of large vesicles known as microvesicles (>1 μm) particles or smaller vesicles named ectosomes (about 100 nm). Alternatively, ECV could be produced by endocytosis of the plasma membrane-forming endosomes. The membrane of these endosomes invaginates toward the lumen resulting in the formation of new vesicles included within a large vesicle that has been named multiple vesicular bodies X. The vesicles inside the multiple vesicular bodies have the same topology of the plasma membrane. When these multiple vesicular bodies fuse with the plasma membrane the vesicle content within the lumen are released outside the cell. ECV derived from this process are known as exosomes [5, 6]. There is extensive evidence that HSP are present within different ECV that is summarized in Table 3.3. The detection of HSP within ECV has primarily been made by mass spectroscopic analysis and, in some cases, confirmed by Western blotting. Since HSP are mainly present in the cytosol, their localization within ECV was assumed to be in the lumen as a result of trapping these proteins during the formation of the vesicles. However, it has been proposed that the composition of ECV is not random but rather a very selective process [6, 115]. Other observations have shown that HSP are located within the membranes of ECV, as in the case of Hsp70 [28, 31, 116], Hsp90 [117], and Hsp60 [118, 119]. The presence of HSP on the membrane (surface) of ECV is important because it may explain a specific interaction with target cells, most likely by a process mediated by surface receptors. On the contrary, the potential biological role of HSP within the lumen of ECV is less evident, which should require the fusion of the vesicles with the plasma membrane or by the burst of ECV liberating the cargo within the extracellular milieu. The presence of HSP on the membrane of ECV has led us to postulate that insertion into the lipid bilayer is the first step in the secretion of this protein [6]. Additional observations have shown that the export of Hsp70 and Hsc70 within ECV was blocked by the reduction of membrane cholesterol levels [79, 120]. Indeed, Hsp70 within ECV was resistant to nonionic detergents, such as Triton X-100, suggesting that the protein is within lipid rafts in the vesicles [28]. In this regard, several studies have shown that Hsp70 is present within lipid rafts of cells [28, 79, 121, 122].

Although the evidence for the active secretion of HSP from living cells is well established, it cannot be ignored that, under other circumstances, HSP are released into the extracellular medium after cell necrosis. Indeed, the concentration of HSP70 released after necrosis could be potentially very high [29]. In this regard, expression of HSP70 has also been observed after ischemia–reperfusion (I/R) injury, which resulted in a necrotic focus [85].

**Table 3.3** Detection of HSP in ECV

| HSP | Cells | Reference |
|---|---|---|
| HSP70 | Dendritic cells | [123] |
| HSP90 | Dendritic cells | [124] |
| HSP70 | Peripheral blood mononuclear cells | [125] |
| HSP70, HSC70, HSP27, HSP90 | B cells | [126] |
| HSC70/HSP70 | Reticulocytes | [127] |
| HSP70, HSP90, Grp78 | Hepatocytes | [128] |
| HSP60 | Cardiac myocytes | [118] |
| HSP70 | Pancreas carcinoma<br>Colon carcinoma | [31] |
| HSP70 | Hepatoblastoma | [28] |
| HSP70 | Thymolymphoma<br>Mammary carcinoma<br>Colon carcinoma | [116] |
| HSP90 | Glioblastoma<br>Fibroblastoma<br>Mammary gland adenocarcinoma | [129] |
| HSP90, HSC70 | Mesothelioma | [130] |
| HSP60 | Cardiac myocytes | [118] |
| HSP60 | Bronchial carcinoma<br>Lung adenocarcinoma<br>Erythroleukemia | [119] |
| HSP70 | *Mycobacterium smegmatis* and *Mycobacterium avium*-infected RAW 264.7 | [131] |

## 3.6 The Stress Observation System

There is clear evidence that cells secrete ECV during normal physiological conditions as well as after stress. We have argued that the composition of ECV reflects the physiological stage of the cell [6]. Thus, constitutive proteins are present in ECV derived from cells under normal conditions, such as CD9 and CD63, which belong to the family of tetraspanin proteins [132]. During stress conditions, ECV should reflect the insult type, such as the presence of stress-inducible HSP. Then, ECV are recognized by other cell types, in particular cells of the immune system, as part of an assessment of the stress conditions. If there is not stress, it is unlikely that there is a response. However, ECV during stress conditions are likely to activate the immune system to prepare a preemptive response directed at avoiding the propagation of the insult (Fig. 3.1). The process of sensing stress via ECV has been termed "Stress Observation System" [6]. Thus, ECV derived from macrophages infected with intracellular pathogens were observed to activate uninfected macrophages by a Toll-like receptor and myeloid differentiation factor 88 dependent mechanism [133]. They also induced polymorphonuclear leukocyte recruitment in lungs after

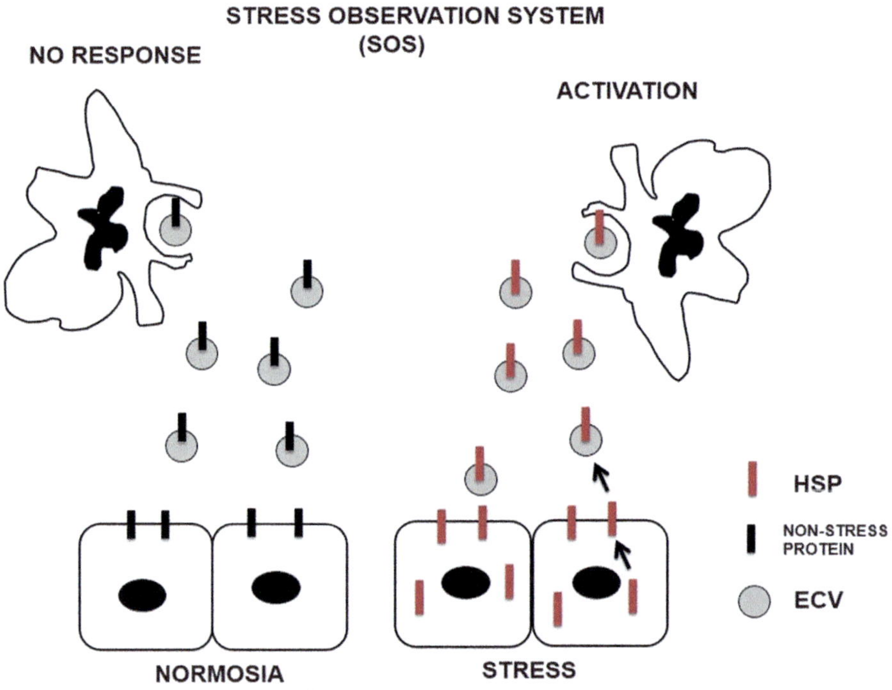


**Fig. 3.1** During normal physiological conditions, cells release ECV containing markers for cellular homeostasis that when captured by immune cells do not trigger any response. However, the composition of these ECV changes after stress, resulting in a signal for the immune system to mount an appropriate response directed at mitigating the insult.

intranasal delivery [133]. Moreover, ECV containing Hsp70 isolated from mycobacteria-infected cells induced an inflammatory response in macrophages [134]. ECV containing Hsp70 on their surface displayed a robust and specific activation of macrophages, which was higher than the same concentration of recombinant Hsp70 in solution [28]. Finally, Hsp70-positive ECV were also found to stimulate the cytotoxic capacity of NK cells [31].

## 3.7 Conclusions

HSP appear to display a different role in the extracellular environment than the well-characterized function as molecular chaperones. Extracellular HSP emerge as new signaling molecules involved in intracellular communication. The presence of extracellular HSP has been detected in biological fluids from individuals suffering from a large number of illnesses. Therefore, they are likely to become biomarkers of various disease conditions. Extracellular HSP are released by many cell types and by at least two main mechanisms, including the passive dissemination after necrosis or an active export process independent on cell death, named the nonclassical secretory pathway. Extracellular HSP come in various flavors. Thus, they can be found in a soluble form within biological fluids, trapped in the lumen of ECV or exposed to the surface of these vesicles in a membrane-bound fashion. Finally, HSP associated with ECV appear to be part of a mechanism directed at both alerting the immune system to the presence of an insult and avoiding the propagation of stress.

**Acknowledgments** This work was supported by the National Institutes of Health, grant number GM R01 09845.

## References

1. Ahmed KA, Xiang J (2011) Mechanisms of cellular communication through intercellular protein transfer. J Cell Mol Med 15:1458–1473
2. Grakoui A, Bromley SK, Sumen C, Davis MM, Shaw AS, Allen PM, Dustin ML (1999) The immunological synapse: a molecular machine controlling T cell activation. Science 285:221–227
3. Davis DM (2007) Intercellular transfer of cell-surface proteins is common and can affect many stages of an immune response. Nat Rev Immunol 7:238–243
4. Arispe N, Doh M, De Maio A (2002) Lipid interaction differentiates the constitutive and stress-induced heat shock proteins Hsc70 and Hsp70. Cell Stress Chaperones 7:330–338
5. Thery C, Ostrowski M, Segura E (2009) Membrane vesicles as conveyors of immune responses. Nat Rev Immunol 9:581–593
6. De Maio A (2011) Extracellular heat shock proteins, cellular export vesicles, and the stress observation system: a form of communication during injury, infection, and cell damage. It is never known how far a controversial finding will go! Dedicated to Ferruccio Ritossa. Cell Stress Chaperones 16:235–249
7. Mulcahy LA, Pink RC, Carter DR (2014) Routes and mechanisms of extracellular vesicle uptake. J Extracell Vesicles 3. https://doi.org/10.3402/jev.v3.24641
8. Ritossa FM (1962) A new puffing pattern induced by temperature shock and DNP in drosophila. Cell Mol Life Sci 18:571–573
9. Tissieres A, Mitchell HK, Tracy UM (1974) Protein synthesis in salivary glands of Drosophila Melanogaster: relation to chromosome puffs. J Mol Biol 84:389–398
10. De Maio A (1999) Heat shock proteins: facts, thoughts, and dreams. Shock 11:1–12
11. Hartl FU, Hayer-Hartl M (2009) Converging concepts of protein folding in vitro and in vivo. Nat Struct Mol Biol 16:574–581
12. Lindquist S, Craig EA (1988) The heat-shock proteins. Annu Rev Genet 22:631–677
13. Becker T, Hartl FU, Wieland F (2002) CD40, an extracellular receptor for binding and uptake of Hsp70-peptide complexes. J Cell Biol 158:1277–1285
14. Kampinga HH, Hageman J, Vos MJ, Kubota H, Tanguay RM, Bruford EA, Cheetham ME, Chen B, Hightower LE (2009) Guidelines for the nomenclature of the human heat shock proteins. Cell Stress Chaperones 14:105–111
15. Tytell M, Greenberg SG, Lasek RJ (1986) Heat shock-like protein is transferred from glia to axon. Brain Res 363:161–164
16. Hightower LE, Guidon PT Jr (1989) Selective release from cultured mammalian cells of heat-shock (stress) proteins that resemble glia-axon transfer proteins. J Cell Physiol 138:257–266
17. Liao DF, Jin ZG, Baas AS, Daum G, Gygi SP, Aebersold R, Berk BC (2000) Purification and identification of secreted oxidative stress-induced factors from vascular smooth muscle cells. J Biol Chem 275:189–196

18. Suzuki S, Kulkarni AB (2010) Extracellular heat shock protein HSP90beta secreted by MG63 osteosarcoma cells inhibits activation of latent TGF-beta1. Biochem Biophys Res Commun 398:525–531
19. Pilzer D, Fishelson Z (2005) Mortalin/GRP75 promotes release of membrane vesicles from immune attacked cells and protection from complement-mediated lysis. Int Immunol 17:1239–1248
20. Delpino A, Castelli M (2002) The 78 kDa glucose-regulated protein (GRP78/BIP) is expressed on the cell membrane, is released into cell culture medium and is also present in human peripheral circulation. Biosci Rep 22:407–420
21. Kern J, Untergasser G, Zenzmaier C, Sarg B, Gastl G, Gunsilius E, Steurer M (2009) GRP-78 secreted by tumor cells blocks the antiangiogenic activity of bortezomib. Blood 114:3960–3967
22. Evdokimovskaya Y, Skarga Y, Vrublevskaya V, Morenkov O (2012) Release of the glucose-regulated protein 94 by baby hamster kidney cells. Cell Biochem Funct 30:558–562
23. Henderson B, Pockley AG (2012) Proteotoxic stress and circulating cell stress proteins in the cardiovascular diseases. Cell Stress Chaperones 17:303–311
24. Nafar F, Williams JB, Mearow KM (2016) Astrocytes release HspB1 in response to amyloid-beta exposure in vitro. J Alzheimers Dis 49:251–263
25. Zuo D, Yu X, Guo C, Yi H, Chen X, Conrad DH, Guo TL, Chen Z, Fisher PB, Subjeck JR, Wang XY (2012) Molecular chaperoning by glucose-regulated protein 170 in the extracellular milieu promotes macrophage-mediated pathogen sensing and innate immunity. FASEB J 26:1493–1505
26. Asea A, Kraeft SK, Kurt-Jones EA, Stevenson MA, Chen LB, Finberg RW, Koo GC, Calderwood SK (2000) HSP70 stimulates cytokine production through a CD14-dependant pathway, demonstrating its dual role as a chaperone and cytokine. Nat Med 6:435–442
27. Vabulas RM, Ahmad-Nejad P, Ghose S, Kirschning CJ, Issels RD, Wagner H (2002) HSP70 as endogenous stimulus of the toll/interleukin-1 receptor signal pathway. J Biol Chem 277:15107–15112
28. Vega VL, Rodriguez-Silva M, Frey T, Gehrmann M, Diaz JC, Steinem C, Multhoff G, Arispe N, De Maio A (2008) Hsp70 translocates into the plasma membrane after stress and is released into the extracellular environment in a membrane-associated form that activates macrophages. J Immunol 180:4299–4307
29. Basu S, Binder RJ, Suto R, Anderson KM, Srivastava PK (2000) Necrotic but not apoptotic cell death releases heat shock proteins, which deliver a partial maturation signal to dendritic cells and activate the NF-kappa B pathway. Int Immunol 12:1539–1546
30. Gastpar R, Gross C, Rossbacher L, Ellwart J, Riegger J, Multhoff G (2004) The cell surface-localized heat shock protein 70 epitope TKD induces migration and cytolytic activity selectively in human NK cells. J Immunol 172:972–980
31. Gastpar R, Gehrmann M, Bausero MA, Asea A, Gross C, Schroeder JA, Multhoff G (2005) Heat shock protein 70 surface-positive tumor exosomes stimulate migratory and cytolytic activity of natural killer cells. Cancer Res 65:5238–5247
32. Aneja R, Odoms K, Dunsmore K, Shanley TP, Wong HR (2006) Extracellular heat shock protein-70 induces endotoxin tolerance in THP-1 cells. J Immunol 177:7184–7192
33. Abboud PA, Lahni PM, Page K, Giuliano JS Jr, Harmon K, Dunsmore KE, Wong HR, Wheeler DS (2008) The role of endogenously produced extracellular hsp72 in mononuclear cell reprogramming. Shock 30:285–292
34. Ortega E, Hinchado MD, Martin-Cordero L, Asea A (2009) The effect of stress-inducible extracellular Hsp72 on human neutrophil chemotaxis: a role during acute intense exercise. Stress 12:240–249
35. Ortega E, Giraldo E, Hinchado MD, Martinez M, Ibanez S, Cidoncha A, Collazos ME, Garcia JJ (2006) Role of Hsp72 and norepinephrine in the moderate exercise-induced stimulation of neutrophils' microbicide capacity. Eur J Appl Physiol 98:250–255
36. Wang R, Kovalchin JT, Muhlenkamp P, Chandawarkar RY (2006) Exogenous heat shock protein 70 binds macrophage lipid raft microdomain and stimulates phagocytosis, processing, and MHC-II presentation of antigens. Blood 107:1636–1642

37. Kovalchin JT, Wang R, Wagh MS, Azoulay J, Sanders M, Chandawarkar RY (2006) In vivo delivery of heat shock protein 70 accelerates wound healing by up-regulating macrophage-mediated phagocytosis. Wound Repair Regen 14:129–137
38. Lopes RL, Borges TJ, Araujo JF, Pinho NG, Bergamin LS, Battastini AM, Muraro SP, Souza AP, Zanin RF, Bonorino C (2014) Extracellular mycobacterial DnaK polarizes macrophages to the M2-like phenotype. PLoS One 9:e113441
39. Borges TJ, Lopes RL, Pinho NG, Machado FD, Souza AP, Bonorino C (2013) Extracellular Hsp70 inhibits pro-inflammatory cytokine production by IL-10 driven down-regulation of C/EBPbeta and C/EBPdelta. Int J Hyperth 29:455–463
40. Takeuchi T, Suzuki M, Fujikake N, Popiel HA, Kikuchi H, Futaki S, Wada K, Nagai Y (2015) Intercellular chaperone transmission via exosomes contributes to maintenance of protein homeostasis at the organismal level. Proc Natl Acad Sci U S A 112:E2497–E2506
41. Pockley AG, Muthana M, Calderwood SK (2008) The dual immunoregulatory roles of stress proteins. Trends Biochem Sci 33:71–79
42. Oura J, Tamura Y, Kamiguchi K, Kutomi G, Sahara H, Torigoe T, Himi T, Sato N (2011) Extracellular heat shock protein 90 plays a role in translocating chaperoned antigen from endosome to proteasome for generating antigenic peptide to be cross-presented by dendritic cells. Int Immunol 23:223–237
43. van Noort JM, Bsibsi M, Nacken P, Gerritsen WH, Amor S (2012) The link between small heat shock proteins and the immune system. Int J Biochem Cell Biol 44:1670–1679
44. Li W, Li Y, Guan S, Fan J, Cheng CF, Bright AM, Chinn C, Chen M, Woodley DT (2007) Extracellular heat shock protein-90alpha: linking hypoxia to skin cell motility and wound healing. EMBO J 26:1221–1233
45. Bhatia A, O'Brien K, Chen M, Woodley DT, Li W (2016) Keratinocyte-secreted heat shock protein-90alpha: leading wound reepithelialization and closure. Adv Wound Care (New Rochelle) 5:176–184
46. Mathur S, Walley KR, Wang Y, Indrambarya T, Boyd JH (2011) Extracellular heat shock protein 70 induces cardiomyocyte inflammation and contractile dysfunction via TLR2. Circ J 75:2445–2452
47. Thuringer D, Berthenet K, Cronier L, Jego G, Solary E, Garrido C (2015) Oncogenic extracellular HSP70 disrupts the gap-junctional coupling between capillary cells. Oncotarget 6:10267–10283
48. Calderwood SK, Gong J, Murshid A (2016) Extracellular HSPs: the complicated roles of extracellular HSPs in immunity. Front Immunol 7:159
49. Binder RJ, Han DK, Srivastava PK (2000) CD91: a receptor for heat shock protein gp96. Nat Immunol 1:151–155
50. Basu S, Binder RJ, Ramalingam T, Srivastava PK (2001) CD91 is a common receptor for heat shock proteins gp96, hsp90, hsp70, and calreticulin. Immunity 14:303–313
51. Wang Y, Kelly CG, Karttunen JT, Whittall T, Lehner PJ, Duncan L, MacAry P, Younson JS, Singh M, Oehlmann W, Cheng G, Bergmeier L, Lehner T (2001) CD40 is a cellular receptor mediating mycobacterial heat shock protein 70 stimulation of CC-chemokines. Immunity 15:971–983
52. Facciponte JG, Wang XY, Subjeck JR (2007) Hsp110 and Grp170, members of the Hsp70 superfamily, bind to scavenger receptor-a and scavenger receptor expressed by endothelial cells-I. Eur J Immunol 37:2268–2279
53. Theriault JR, Adachi H, Calderwood SK (2006) Role of scavenger receptors in the binding and internalization of heat shock protein 70. J Immunol 177:8604–8611
54. Yang S, Vigerust DJ, Shepherd VL (2013) Interaction of members of the heat shock protein-70 family with the macrophage mannose receptor. J Leukoc Biol 93:529–536
55. Alard JE, Hillion S, Guillevin L, Saraux A, Pers JO, Youinou P, Jamin C (2011) Autoantibodies to endothelial cell surface ATP synthase, the endogenous receptor for hsp60, might play a pathogenic role in vasculatides. PLoS One 6:e14654
56. Fong JJ, Sreedhara K, Deng L, Varki NM, Angata T, Liu Q, Nizet V, Varki A (2015) Immunomodulatory activity of extracellular Hsp70 mediated via paired receptors Siglec-5 and Siglec-14. EMBO J 34:2775–2788

57. Gehrmann M, Liebisch G, Schmitz G, Anderson R, Steinem C, De Maio A, Pockley G, Multhoff G (2008) Tumor-specific Hsp70 plasma membrane localization is enabled by the glycosphingolipid Gb3. PLoS One 3:e1925
58. Sugawara S, Kawano T, Omoto T, Hosono M, Tatsuta T, Nitta K (2009) Binding of Silurus Asotus lectin to Gb3 on Raji cells causes disappearance of membrane-bound form of HSP70. Biochim Biophys Acta 1790:101–109
59. Arispe N, Doh M, Simakova O, Kurganov B, De Maio A (2004) Hsc70 and Hsp70 interact with phosphatidylserine on the surface of PC12 cells resulting in a decrease of viability. FASEB J 18:1636–1645
60. Schilling D, Gehrmann M, Steinem C, De Maio A, Pockley AG, Abend M, Molls M, Multhoff G (2009) Binding of heat shock protein 70 to extracellular phosphatidylserine promotes killing of normoxic and hypoxic tumor cells. FASEB J 23:2467–2477
61. Nylandsted J, Gyrd-Hansen M, Danielewicz A, Fehrenbacher N, Lademann U, Hoyer-Hansen M, Weber E, Multhoff G, Rohde M, Jaattela M (2004) Heat shock protein 70 promotes cell survival by inhibiting lysosomal membrane permeabilization. J Exp Med 200:425–435
62. Pockley AG, Shepherd J, Corton JM (1998) Detection of heat shock protein 70 (Hsp70) and anti-Hsp70 antibodies in the serum of normal individuals. Immunol Investig 27:367–377
63. Zhu J, Quyyumi AA, Wu H, Csako G, Rott D, Zalles-Ganley A, Ogunmakinwa J, Halcox J, Epstein SE (2003) Increased serum levels of heat shock protein 70 are associated with low risk of coronary artery disease. Arterioscler Thromb Vasc Biol 23:1055–1059
64. Zhang X, He M, Cheng L, Chen Y, Zhou L, Zeng H, Pockley AG, Hu FB, Wu T (2008) Elevated heat shock protein 60 levels are associated with higher risk of coronary heart disease in Chinese. Circulation 118:2687–2693
65. Zhang X, Xu Z, Zhou L, Chen Y, He M, Cheng L, Hu FB, Tanguay RM, Wu T (2010) Plasma levels of Hsp70 and anti-Hsp70 antibody predict risk of acute coronary syndrome. Cell Stress Chaperones 15:675–686
66. Pockley AG, Wu R, Lemne C, Kiessling R, de Faire U, Frostegard J (2000) Circulating heat shock protein 60 is associated with early cardiovascular disease. Hypertension 36:303–307
67. Lewthwaite J, Owen N, Coates A, Henderson B, Steptoe A (2002) Circulating human heat shock protein 60 in the plasma of British civil servants: relationship to physiological and psychosocial stress. Circulation 106:196–201
68. Krishnamurthy K, Kanagasabai R, Druhan LJ, Ilangovan G (2012) Heat shock protein 25-enriched plasma transfusion preconditions the heart against doxorubicin-induced dilated cardiomyopathy in mice. J Pharmacol Exp Ther 341:829–839
69. Gehrmann M, Cervello M, Montalto G, Cappello F, Gulino A, Knape C, Specht HM, Multhoff G (2014) Heat shock protein 70 serum levels differ significantly in patients with chronic hepatitis, liver cirrhosis, and hepatocellular carcinoma. Front Immunol 5:307
70. Liao WC, Wu MS, Wang HP, Tien YW, Lin JT (2009) Serum heat shock protein 27 is increased in chronic pancreatitis and pancreatic carcinoma. Pancreas 38:422–426
71. Melle C, Ernst G, Escher N, Hartmann D, Schimmel B, Bleul A, Thieme H, Kaufmann R, Felix K, Friess HM, Settmacher U, Hommann M, Richter KK, Daffner W, Taubig H, Manger T, Claussen U, von Eggeling F (2007) Protein profiling of microdissected pancreas carcinoma and identification of HSP27 as a potential serum marker. Clin Chem 53:629–635
72. Yuan J, Dunn P, Martinus RD (2011) Detection of Hsp60 in saliva and serum from type 2 diabetic and non-diabetic control subjects. Cell Stress Chaperones 16:689–693
73. Oglesbee MJ, Herdman AV, Passmore GG, Hoffman WH (2005) Diabetic ketoacidosis increases extracellular levels of the major inducible 70-kDa heat shock protein. Clin Biochem 38:900–904
74. Genth-Zotz S, Bolger AP, Kalra PR, von Haehling S, Doehner W, Coats AJ, Volk HD, Anker SD (2004) Heat shock protein 70 in patients with chronic heart failure: relation to disease severity and survival. Int J Cardiol 96:397–401
75. Dybdahl B, Slordahl SA, Waage A, Kierulf P, Espevik T, Sundan A (2005) Myocardial ischaemia and the inflammatory response: release of heat shock protein 70 after myocardial infarction. Heart 91:299–304

76. Azuma K, Shichijo S, Takedatsu H, Komatsu N, Sawamizu H, Itoh K (2003) Heat shock cognate protein 70 encodes antigenic epitopes recognised by HLA-B4601-restricted cytotoxic T lymphocytes from cancer patients. Br J Cancer 89:1079–1085
77. Faure O, Graff-Dubois S, Bretaudeau L, Derre L, Gross DA, Alves PM, Cornet S, Duffour MT, Chouaib S, Miconnet I, Gregoire M, Jotereau F, Lemonnier FA, Abastado JP, Kosmatopoulos K (2004) Inducible Hsp70 as target of anticancer immunotherapy: identification of HLA-A*0201-restricted epitopes. Int J Cancer 108:863–870
78. Wu FH, Yuan Y, Li D, Liao SJ, Yan B, Wei JJ, Zhou YH, Zhu JH, Zhang GM, Feng ZH (2012) Extracellular HSPA1A promotes the growth of hepatocarcinoma by augmenting tumor cell proliferation and apoptosis-resistance. Cancer Lett 317:157–164
79. Hunter-Lavin C, Davies EL, Bacelar MM, Marshall MJ, Andrew SM, Williams JH (2004) Hsp70 release from peripheral blood mononuclear cells. Biochem Biophys Res Commun 324:511–517
80. Pittet JF, Lee H, Morabito D, Howard MB, Welch WJ, Mackersie RC (2002) Serum levels of Hsp 72 measured early after trauma correlate with survival. J Trauma 52:611–617. discussion 617
81. Ziegler TR, Ogden LG, Singleton KD, Luo M, Fernandez-Estivariz C, Griffith DP, Galloway JR, Wischmeyer PE (2005) Parenteral glutamine increases serum heat shock protein 70 in critically ill patients. Intensive Care Med 31:1079–1086
82. Ganter MT, Ware LB, Howard M, Roux J, Gartland B, Matthay MA, Fleshner M, Pittet JF (2006) Extracellular heat shock protein 72 is a marker of the stress protein response in acute lung injury. Am J Physiol Lung Cell Mol Physiol 291:L354–L361
83. Flohe SB, Bangen JM, Flohe S, Agrawal H, Bergmann K, Schade FU (2007) Origin of immunomodulation after soft tissue trauma: potential involvement of extracellular heat-shock proteins. Shock 27:494–502
84. Hecker JG, McGarvey M (2011) Heat shock proteins as biomarkers for the rapid detection of brain and spinal cord ischemia: a review and comparison to other methods of detection in thoracic aneurysm repair. Cell Stress Chaperones 16:119–131
85. De Maio A, Vazquez D (2013) Extracellular heat shock proteins: a new location, a new function. Shock 40:239–246
86. Njemini R, Lambert M, Demanet C, Mets T (2003) Elevated serum heat-shock protein 70 levels in patients with acute infection: use of an optimized enzyme-linked immunosorbent assay. Scand J Immunol 58:664–669
87. Osorio-Caballero M, Perdigon-Palacio C, Garcia-Lopez G, Flores-Herrera O, Olvera-Sanchez S, Morales-Mendez I, Sosa-Gonzalez I, Acevedo JF, Guzman-Grenfell AM, Molina-Hernandez A, Diaz NF, Flores-Herrera H (2015) Escherichia coli-induced temporal and differential secretion of heat-shock protein 70 and interleukin-1beta by human fetal membranes in a two-compartment culture system. Placenta 36:262–269
88. Molvarec A, Prohaszka Z, Nagy B, Szalay J, Fust G, Karadi I, Rigo J Jr (2006) Association of elevated serum heat-shock protein 70 concentration with transient hypertension of pregnancy, preeclampsia and superimposed preeclampsia: a case-control study. J Hum Hypertens 20:780–786
89. Molvarec A, Tamasi L, Losonczy G, Madach K, Prohaszka Z, Rigo J Jr (2010) Circulating heat shock protein 70 (HSPA1A) in normal and pathological pregnancies. Cell Stress Chaperones 15:237–247
90. Asea A, Jean-Pierre C, Kaur P, Rao P, Linhares IM, Skupski D, Witkin SS (2008) Heat shock protein-containing exosomes in mid-trimester amniotic fluids. J Reprod Immunol 79:12–17
91. Walsh RC, Koukoulas I, Garnham A, Moseley PL, Hargreaves M, Febbraio MA (2001) Exercise increases serum Hsp72 in humans. Cell Stress Chaperones 6:386–393
92. Febbraio MA, Ott P, Nielsen HB, Steensberg A, Keller C, Krustrup P, Secher NH, Pedersen BK (2002) Exercise induces hepatosplanchnic release of heat shock protein 72 in humans. J Physiol 544:957–962
93. Periard JD, Ruell P, Caillaud C, Thompson MW (2012) Plasma Hsp72 (HSPA1A) and Hsp27 (HSPB1) expression under heat stress: influence of exercise intensity. Cell Stress Chaperones 17:375–383

94. Santos TM, Sinzato YK, Gallego FQ, Iessi IL, Volpato GT, Dallaqua B, Damasceno DC (2015) Extracellular HSP70 levels in diabetic environment in rats. Cell Stress Chaperones 20:595–603
95. Tsai TN, Lee TY, Liu MS, Chuang IC, Lu MC, Dong HP, Lue SI, Yang RC (2015) Release of endogenous heat shock protein 72 on the survival of sepsis in rats. J Surg Res 198:165–174
96. Wilhelmus MM, Boelens WC, Otte-Holler I, Kamps B, de Waal RM, Verbeek MM (2006) Small heat shock proteins inhibit amyloid-beta protein aggregation and cerebrovascular amyloid-beta protein toxicity. Brain Res 1089:67–78
97. Wilhelmus MM, Boelens WC, Otte-Holler I, Kamps B, Kusters B, Maat-Schieman ML, de Waal RM, Verbeek MM (2006) Small heat shock protein HspB8: its distribution in Alzheimer's disease brains and its inhibition of amyloid-beta protein aggregation and cerebrovascular amyloid-beta toxicity. Acta Neuropathol 111:139–149
98. Evans CG, Wisen S, Gestwicki JE (2006) Heat shock proteins 70 and 90 inhibit early stages of amyloid beta-(1-42) aggregation in vitro. J Biol Chem 281:33182–33191
99. Carnini A, Scott LO, Ahrendt E, Proft J, Winkfein RJ, Kim SW, Colicos MA, Braun JE (2012) Cell line specific modulation of extracellular abeta42 by Hsp40. PLoS One 7:e37755
100. Luo X, Tao L, Lin P, Mo X, Chen H (2012) Extracellular heat shock protein 72 protects schwann cells from hydrogen peroxide-induced apoptosis. J Neurosci Res 90:1261–1269
101. Guzhova I, Kislyakova K, Moskaliova O, Fridlanskaya I, Tytell M, Cheetham M, Margulis B (2001) In vitro studies show that Hsp70 can be released by glia and that exogenous Hsp70 can enhance neuronal stress tolerance. Brain Res 914:66–73
102. Zhu Z, Li R, Stricker R, Reiser G (2015) Extracellular alpha-crystallin protects astrocytes from cell death through activation of MAPK, PI3K/Akt signaling pathway and blockade of ROS release from mitochondria. Brain Res 1620:17–28
103. Nickel W, Seedorf M (2008) Unconventional mechanisms of protein transport to the cell surface of eukaryotic cells. Annu Rev Cell Dev Biol 24:287–308
104. Wimley WC, Hristova K, Ladokhin AS, Silvestro L, Axelsen PH, White SH (1998) Folding of beta-sheet membrane proteins: a hydrophobic hexapeptide model. J Mol Biol 277:1091–1110
105. Arispe N, De Maio A (2000) ATP and ADP modulate a cation channel formed by Hsc70 in acidic phospholipid membranes. J Biol Chem 275:30839–30843
106. Macazo FC, White RJ (2014) Monitoring charge flux to quantify unusual ligand-induced ion channel activity for use in biological nanopore-based sensors. Anal Chem 86:5519–5525
107. Armijo G, Okerblom J, Cauvi DM, Lopez V, Schlamadinger DE, Kim J, Arispe N, De Maio A (2014) Interaction of heat shock protein 70 with membranes depends on the lipid environment. Cell Stress Chaperones 19:877–886
108. McCallister C, Kdeiss B, Nikolaidis N (2016) Biochemical characterization of the interaction between HspA1A and phospholipids. Cell Stress Chaperones 21:41–53
109. Lopez V, Cauvi DM, Arispe N, De Maio A (2016) Bacterial Hsp70 (DnaK) and mammalian Hsp70 interact differently with lipid membranes. Cell Stress Chaperones 21:609–616
110. Multhoff G, Hightower LE (1996) Cell surface expression of heat shock proteins and the immune response. Cell Stress Chaperones 1:167–176
111. Multhoff G (2007) Heat shock protein 70 (Hsp70): membrane location, export and immunological relevance. Methods 43:229–237
112. Mambula SS, Calderwood SK (2006) Heat shock protein 70 is secreted from tumor cells by a nonclassical pathway involving lysosomal endosomes. J Immunol 177:7849–7857
113. Andrei C, Dazzi C, Lotti L, Torrisi MR, Chimini G, Rubartelli A (1999) The secretory route of the leaderless protein interleukin 1beta involves exocytosis of endolysosome-related vesicles. Mol Biol Cell 10:1463–1475
114. Evdonin AL, Martynova MG, Bystrova OA, Guzhova IV, Margulis BA, Medvedeva ND (2006) The release of Hsp70 from A431 carcinoma cells is mediated by secretory-like granules. Eur J Cell Biol 85:443–455
115. Janas T, Janas MM, Sapon K, Janas T (2015) Mechanisms of RNA loading into exosomes. FEBS Lett 589:1391–1398

116. Chalmin F, Ladoire S, Mignot G, Vincent J, Bruchard M, Remy-Martin JP, Boireau W, Rouleau A, Simon B, Lanneau D, De Thonel A, Multhoff G, Hamman A, Martin F, Chauffert B, Solary E, Zitvogel L, Garrido C, Ryffel B, Borg C, Apetoh L, Rebe C, Ghiringhelli F (2010) Membrane-associated Hsp72 from tumor-derived exosomes mediates STAT3-dependent immunosuppressive function of mouse and human myeloid-derived suppressor cells. J Clin Invest 120:457–471
117. Li W, Sahu D, Tsen F (2012) Secreted heat shock protein-90 (Hsp90) in wound healing and cancer. Biochim Biophys Acta 1823:730–741
118. Gupta S, Knowlton AA (2007) HSP60 trafficking in adult cardiac myocytes: role of the exosomal pathway. Am J Physiol Heart Circ Physiol 292:H3052–H3056
119. Merendino AM, Bucchieri F, Campanella C, Marciano V, Ribbene A, David S, Zummo G, Burgio G, Corona DF, Conway de Macario E, Macario AJ, Cappello F (2010) Hsp60 is actively secreted by human tumor cells. PLoS One 5:e9247
120. Evdokimovskaya Y, Skarga Y, Vrublevskaya V, Morenkov O (2010) Secretion of the heat shock proteins HSP70 and HSC70 by baby hamster kidney (BHK-21) cells. Cell Biol Int 34:985–990
121. Broquet AH, Thomas G, Masliah J, Trugnan G, Bachelet M (2003) Expression of the molecular chaperone Hsp70 in detergent-resistant microdomains correlates with its membrane delivery and release. J Biol Chem 278:21601–21606
122. Chen S, Bawa D, Besshoh S, Gurd JW, Brown IR (2005) Association of heat shock proteins and neuronal membrane components with lipid rafts from the rat brain. J Neurosci Res 81:522–529
123. Thery C, Regnault A, Garin J, Wolfers J, Zitvogel L, Ricciardi-Castagnoli P, Raposo G, Amigorena S (1999) Molecular characterization of dendritic cell-derived exosomes. Selective accumulation of the heat shock protein hsc73. J Cell Biol 147:599–610
124. Chaput N, Flament C, Viaud S, Taieb J, Roux S, Spatz A, Andre F, LePecq JB, Boussac M, Garin J, Amigorena S, Thery C, Zitvogel L (2006) Dendritic cell derived-exosomes: biology and clinical implementations. J Leukoc Biol 80:471–478
125. Lancaster GI, Febbraio MA (2005) Exosome-dependent trafficking of HSP70: a novel secretory pathway for cellular stress proteins. J Biol Chem 280:23349–23355
126. Clayton A, Turkes A, Navabi H, Mason MD, Tabi Z (2005) Induction of heat shock proteins in B-cell exosomes. J Cell Sci 118:3631–3638
127. Mathew A, Bell A, Johnstone RM (1995) Hsp-70 is closely associated with the transferrin receptor in exosomes from maturing reticulocytes. Biochem J 308(Pt 3):823–830
128. Conde-Vancells J, Rodriguez-Suarez E, Embade N, Gil D, Matthiesen R, Valle M, Elortza F, Lu SC, Mato JM, Falcon-Perez JM (2008) Characterization and comprehensive proteome profiling of exosomes secreted by hepatocytes. J Proteome Res 7:5157–5166
129. McCready J, Sims JD, Chan D, Jay DG (2010) Secretion of extracellular hsp90alpha via exosomes increases cancer cell motility: a role for plasminogen activation. BMC Cancer 10:294
130. Hegmans JP, Bard MP, Hemmes A, Luider TM, Kleijmeer MJ, Prins JB, Zitvogel L, Burgers SA, Hoogsteden HC, Lambrecht BN (2004) Proteomic analysis of exosomes secreted by human mesothelioma cells. Am J Pathol 164:1807–1815
131. Anand PK, Anand E, Bleck CK, Anes E, Griffiths G (2010) Exosomal Hsp70 induces a proinflammatory response to foreign particles including mycobacteria. PLoS One 5:e10136
132. Maecker HT, Todd SC, Levy S (1997) The tetraspanin superfamily: molecular facilitators. FASEB J 11:428–442
133. Bhatnagar S, Shinagawa K, Castellino FJ, Schorey JS (2007) Exosomes released from macrophages infected with intracellular pathogens stimulate a proinflammatory response in vitro and in vivo. Blood 110:3234–3244
134. O'Neill HC, Quah BJ (2008) Exosomes secreted by bacterially infected macrophages are proinflammatory. Sci Signal 1:pe8

# Part III
# Regulation of Immune Responses by Extracellular HSPs

# Chapter 4
# The Heat Shock Protein-CD91 Pathway and Tumor Immunosurveillance

**Robert J. Binder**

**Abstract** The intracellular functions of HSPs have been well studied and delineate a clear role in the unfolded protein response. The functions of extracellular HSPs are only beginning to be appreciated. Specifically, extracellular localization of HSPs endorses the initiation of immune responses against aberrant cells. This chapter examines the role of extracellular HSPs, and the receptor CD91, in immunosurveillance of cancers. Although the concept of cancer immunosurveillance was described over 100 years ago, a molecular description of how the immune responses is initiated has been lacking. Incorporating the HSP-CD91 pathway into cancer immunosurveillance provides the first mechanism of how immune responses are primed.

**Keywords** Dendritic cell · Chaperone · Tumor immunity · T regs

## 4.1 Heat Shock Proteins as Chaperones of Macromolecules

Heat shock proteins have long been known for their function as chaperones within cells, where they assist proteins and polypeptides fold into their native, most stable configurations [1, 2]. Many HSPs are inducible by cellular stress [1], a condition where there is a heightened requirement for chaperone function. However, several other HSPs are constitutively expressed. Recently, the chaperone function of HSPs has been shown to be required for transport of other macromolecules. These macromolecules include peptides derived from homeostatic protein turnover [3–9]. This latter function has been implicated in several immunological processes and pathways. For example, peptides in the MHC I processing and presentation pathway are shuttled by HSPs in the cytosol and endoplasmic reticulum [4–6]. Although the normal expression pattern of HSPs is solely intracellular, under certain pathological conditions, HSPs can be found in the extracellular environment, free as a diffusible

R. J. Binder (✉)
Department of Immunology, University of Pittsburgh School of Medicine, Pittsburgh, PA, USA
e-mail: rjb42@pitt.edu

 
R. J. Binder, P. K. Srivastava (eds.), *Heat Shock Proteins in the Immune System*,
https://doi.org/10.1007/978-3-319-69042-1_4

soluble protein [10, 11], as part of the extracellular matrix [11] or on the membrane of cells [12]. Infection by pathogens, hostile cancer microenvironments, and inflammation associated with these events, very frequently, if not always, include cell death which leads to passive release of these abundant chaperones [10]. As described below, the chaperone function of HSPs is critical to its role in the immune system.

## 4.2 Immune Responses Elicited by Extracellular HSPs

In the extracellular environment, select HSPs have been shown to elicit immune responses of diverse nature [13–25]. This remarkable property was first observed when gp96 was isolated as the immunogenic entity of tumor cells [13]. In that pioneering study, when mice where immunized with gp96 preparations derived from a tumor, they became resistant to a subsequent challenge of that tumor. This phenomenon has been replicated for hsp70 [14], hsp90 [15], calreticulin [16], grp170 [17], and hsp110 [17], the major chaperones of cells. HSPs chaperone peptides [3–9], and when isolated from tumor cells, the peptide repertoire includes tumor antigens [8, 9, 13–17]. In other words, purified HSP-peptide complexes represent the antigenic fingerprint of the cell from which they are isolated. This has been empirically tested. In several antigenically defined systems, HSPs have been shown to be associated with antigens that ultimately get presented by MHC I and MHC II molecules thereby dictating T cell specificities of the immune response. These systems include HSPs isolated from tumors [8, 9, 26–28], infected cells [29–34], allo-MHC cells [35, 36], and cells expressing model antigens [35–38]. In studies where crystal structures of HSPs have been resolved, peptide binding pockets have been clearly identified [39–41]. Over two decades of work has elucidated the major immunological mechanisms through which HSPs prime immune responses. These mechanisms are dependent on the ability of HSP to bind cell surface receptors on antigen-presenting cells (APC) [42]. In the extracellular environment, HSPs engage a cell surface receptor, CD91, which is expressed by most APCs [20, 43–55]. On conventional dendritic cells, CD91 acts as an endocytic receptor to internalize HSP-peptide complexes [43–46]. Several other cell surface receptors for the immunogenic HSPs have been proposed and are discussed elsewhere [42]. Following CD91-dependent endocytosis, the HSP-peptide complexes are processed, and the peptides enter the pathways for antigen presentation for MHC I [43, 44, 47] or MHC II [45, 48] of the APC. CD91 also acts as a signaling receptor [20]. Upon engagement by HSPs, various signaling and transcription factors are activated following phosphorylation of the CD91 cytoplasmic chain, leading to production and secretion of cytokines and upregulation of co-stimulatory molecules [10, 20, 51]. On conventional dendritic cells, the signaling pathways and outcomes are responsible for and supportive of Th1 responses and subsequent HSP-mediated rejection of tumors and pathogens following vaccination. Interestingly CD91 is expressed by hematopoietic cells of both myeloid and lymphoid origin including macrophages and a variety of DC subsets [52–55]. When HSPs are in the extracellular environment, HSPs can engage

CD91 on any cell in that microenvironment or can drain to lymph nodes and engage (additional) cells at this distal site [52]. Using fluorescent tags, HSPs were shown to engage cDCs in vivo at doses capable of priming Th1 responses [52]. However, increasing amounts of HSPs will engage additional cells, including pDCs [53, 91]. The exact phenotype of the immunological responses is determined by the $CD91^{+}$ APC engaged by the extracellular HSP. For example, pDCs engage extracellular HSPs but do not cross-present HSP-chaperoned peptides nor upregulate B7 or CD40 [53], promoting an immune-regulatory phenotype characterized by T reg [91]. These responses have been harnessed for immunotherapy of autoimmune disease and amelioration of tissue allograft acceptance [21–23]. Engagement of cDCs by the same HSPs promote Th1 response that reject tumors [43, 44, 47, 52]. The influence of other tumor-secreted molecules, besides HSPs, in the immediate microenvironment potentially also plays a role in the resulting immune response [20]. Molecules like HMGB1 [56], dsDNA [57], and cytokines [20] have been shown to be immunologically important and could complement or antagonize the responses emanating from the HSP-APC interaction. For example, tumor-secreted TGF-β synergizes with HSP/CD91-dependent IL-6 and TNF-α released from APCs to prime Th17 responses [20]. The resulting immunological response elicited by extracellular HSPs will be dependent on the influence of local APCs on cross-priming by cDCs in the draining lymph node. Many of these mechanisms, while demonstrated in murine models, also hold true in the human setting [58, 59].

## 4.3 Extracellular HSPs as the Molecular Signature for Immunological Responsiveness

A majority of the findings described above have been performed in a vaccination setting where purified HSPs are administered to rodents or humans [13–19, 21–24]. However in studies examining HSPs released from cells in situ, the same stimulation of APCs can be observed [52, 60]. Under pathological conditions and cell death, HSPs are released from cells and delivered to the extracellular environment [10–12]. Mechanisms of active secretion of HSPs have also been put forward to explain the extracellular presence of HSPs, but these are not fully elucidated [61]. Since HSPs contain no consensus sequences for such cellular trafficking and secretion, it is hard to conceive the cell biology comprising such a pathway, especially for the cytosolic HSPs. Thus, a passive release mechanism, when membrane integrity is compromised, appears more likely. Examples of these pathological conditions leading to HSP release include cellular infection by bacteria and viruses, cancer, trauma, and associated inflammation. Collectively, HSPs are the most abundant proteins in cells accounting for >5% of the proteome [1]. Thus, they are ideal indicators to the immune system of cellular aberrancy. There are now six key HSPs known to be rapidly recognized by the APCs [13–17] via cell surface receptor(s). The surprising discovery of the HSP receptor expressed on APCs afforded a molecular description of these immunological mechanisms [43]. Since the receptor(s) offers a significant degree of specificity for recognition of intracellular content, they become key players in the immune system, allowing HSPs to be critical initiators *and* mediators of resulting immune responses. Following the initiation of antigen-specific immune responses against cancers or pathogen-infected cells, extracellular HSPs exacerbate existing inflammatory conditions or suppress ongoing immunity [60]. There is currently a well-developed picture on the cross-presentation of HSP-chaperoned peptides to which T cells are primed and pathways which lead to the release of cytokines, including the pro-inflammatory IL-1, IL-6, and TNF-α [10, 20]. Thus, extracellular HSPs have been implicated in the etiology, progression, and/or resolution of several diseases including cancer and rheumatoid arthritis [60, 62, 63]. In rheumatoid arthritis, the presence of extracellular hsp70 and gp96 in synovial fluid of inflamed joints has been shown to stimulate local APCs which release pro-inflammatory cytokines. These events constitute a cycle of tissue destruction, increased release of HSPs, and increased inflammation [62, 63]. Recognition of endogenous molecules (HSPs) by their respective receptors can be compared on many levels to the recognition of pathogen-associated molecular patterns (PAMPs) by pattern recognition receptors (PRRs) [64].

## 4.4 The Role of Extracellular HSPs in Tumor Immunosurveillance

The original concept of immunosurveillance was that the immune system recognized aberrant cells and eliminated them before progression to cancer occurred [65–67]. We now know that priming of T cell and NK cell immunity is necessary for rejection of aberrant cells. In the absence of such immunity in mice [68, 69] or in humans [70], achieved by the loss of these immune cells themselves or their effector molecules, multiple and frequent tumors arise. The tumors that arise under these immune-compromised conditions are less edited compared to tumors from wild-type mice [68, 71]. The literature, however, until recently, failed to reconcile two issues. The first pertains to the miniscule amount of antigen available for priming T cell responses at the very earliest stages of nascent tumor development [72, 73]. The realization that most tumor *rejection* antigens are unique and derived from mutated proteins [74–77] predicts that antigen levels in (emerging) tumors (and the quantity available for cross-presentation) are minute and, as a soluble protein, have indeed been shown to be insufficient for cross-priming of T cell responses [72, 73]. Yet, T cell responses are easily measurable at these early time points of tumorigenesis (e.g. [78, 79]). Mechanisms of antigen transfer and cross-presentation described for other systems [80–87] where antigen is abundant or supraphysiological are not justifiable for nascent emerging tumors. Thus, a super-efficient mechanism must exist for antigen cross-presentation in this setting [88]. Experimental evidence shows that these quantitative restrictions are satisfied if one invokes the HSP-peptide complexes

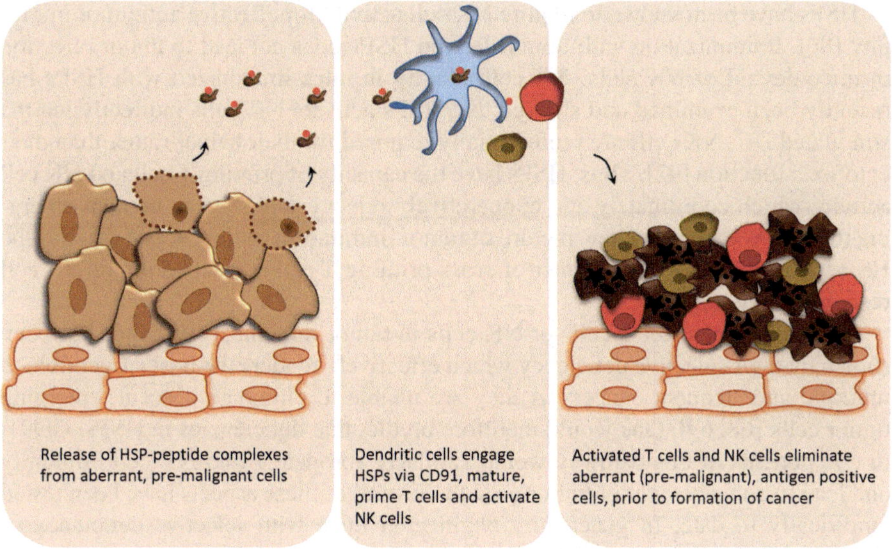


**Fig. 4.1** HSPs prime immune responses responsible for eradication of premalignant cells. HSPs released from aberrant, membrane-compromised cells engage dendritic cells in the draining lymph nodes via the receptor CD91. Dendritic cells mature and cross-present HSP-chaperoned antigens to T cells. T cells are primed and NK cells are activated by these DCs. Activated effector cells eliminate aberrant, premalignant cells prior to formation of cancer

released by tumor cells as a mechanism of antigen transfer [60, 88]. When tumor antigen levels are low, peptides derived from tumor antigens and chaperoned by HSP are efficiently cross-presented by APCs, a system that is dependent on CD91 expressed on APCs [43, 44]. One microgram of total immunogenic HSP (an amount that will be present in <10,000 cells) will chaperone approximately a nanogram of a specific antigenic/mutated peptide. This amount of antigen is sufficient for cross-priming only when chaperoned by the HSP (Fig. 4.1).

The second issue relates to the stimuli in the setting of nascent, emerging tumors that results in co-stimulation for T cell priming. Over the millennia, the immune system has evolved to recognize PAMPs associated with pathogens but are necessarily absent in the host [64]. PAMPs generate co-stimulation and cytokines for T cell priming through well-defined pathways. Nascent tumors lack PAMPs and so will be unable to elicit co-stimulation via PRRs. Interestingly, a very short list of molecules of *host* origin, typically called DAMPs, can do so [10, 20, 56, 57]. HSPs are the prototypical DAMPs, the first group of host molecules found to stimulate DCs to release cytokines, upregulate co-stimulatory molecule expression [10], and prime immune responses [13]. The HSP/DAMP receptor, CD91, channels intracellular signals to achieve this, and the co-stimulation provided by APCs has been well defined [10, 20]. Thus, tumor-derived HSP-peptide complexes are a single entity with the capacity of priming robust antigen-specific T cell responses, without the requirement of additional adjuvanticity or antigen.

HSPs have been known to require NK cell activity for effective antitumor immunity [89]. Immunization with tumor-derived HSPs does not lead to tumor rejection in mice devoid of NK cells. NK cell activity in mice immunized with HSPs has recently been examined and showed that HSPs activate NK cells indirectly via the stimulated DC. NK cells are preferentially required for their helper rather than their cytotoxic function [92]. Thus, HSPs have the capacity of priming T cell and NK cell activity which coordinately and cooperatively reject established or nascent emerging tumors. We present a new picture of tumor immunosurveillance, one that has the HSP-CD91 pathway at the center of cross-priming T cell and activation of NK cell responses (Fig. 4.1).

The requirement for T cells or NK cells in tumor immunosurveillance has been shown by their selective deficiency which effectively renders the host susceptible to multiple and frequent cancers as they are unable to eliminate nascent, emerging tumor cells [68, 69]. One would therefore predict that deficiencies in HSPs, CD91, or components of this pathway would similarly abrogate T and NK cell immunity and lead to enhancement of tumor growth. Several of these aspects have been tested empirically to date. In genetically engineered mice with selective deficiency of CD91 in APCs, HSPs are unable to cross-present chaperoned peptides and stimulate co-stimulation [60]. These mice thus fail to mount tumor-specific T cells and control tumor growth. The immunogenic HSPs play redundant roles in cross-priming, and since their collective deletion in mice is not feasible, the alternative experiment with deficiencies in HSPs is technically challenging. However, when HSPs are collectively deleted in tumor cell lysates, the resulting lysates are incapable of priming tumor-specific immunity, even though they contain soluble tumor antigen [72]. These results cumulatively point to the HSP-CD91 pathway as essential for priming immune responses against tumors and for tumor immunosurveillance. While other DAMPs such as HMGB1 and dsDNA may contribute additional cytokines or co-stimulation through APCs, they do not appear to be essential for tumor immunity, but may influence ongoing responses.

## 4.5 Conclusion

Defining the role of tumor-derived HSPs and CD91 in tumor immunosurveillance is still in its infancy, but the current experimental evidence supporting this premise is significant. There is now an original molecular mechanism as to how immune response, constituting CTL and NK cell activity, is initiated against a nascent, emerging tumor and how this leads to rejection of tumors. The evidence supporting this model also fulfils the quantitative restrictions defined by the scarcity of the tumor antigens. In a tumor microenvironment, with release of multiple HSPs and in the presence of several different APC populations, the immune response is fluid but can be of the Th1 type for tumor rejection. This response may also be fine-tuned by other factors such as additional DAMPs or molecules associated with DNA damage

[90]. The clinical implications of HSP-mediated immunogenicity are currently being investigated.

## References


1. Lindquist S, Craig EA (1988) The heat-shock proteins. Annu Rev Genet 22:631–677
2. Jeng W, Lee S, Sung N, Lee J, Tsai FT (2015) Molecular chaperones: guardians of the proteome in normal and disease states. F1000Res 15:4
3. Li Z, Srivastava PK (1993) Tumor rejection antigen gp96/grp94 is an ATPase: implications for protein folding and antigen presentation. EMBO J 12:3143–3151
4. Kunisawa J, Shastri N (2006) Hsp90alpha chaperones large C-terminally extended proteolytic intermediates in the MHC class I antigen processing pathway. Immunity 24:523–534
5. Srivastava PK, Udono H, Blachere NE, Li Z (1994) Heat shock proteins transfer peptides during antigen processing and CTL priming. Immunogenetics 39:93–98
6. Callahan MK, Garg M, Srivastava PK (2008) Heat-shock protein 90 associates with N-terminal extended peptides and is required for direct and indirect antigen presentation. Proc Natl Acad Sci U S A 105:1662–1667
7. Demine R, Walden P (2005) Testing the role of gp96 as peptide chaperone in antigen processing. J Biol Chem 280:17573–17578
8. Grossmann ME, Madden BJ, Gao F, Pang YP, Carpenter JE, McCormick D, Young CY (2004) Proteomics shows Hsp70 does not bind peptide sequences indiscriminately in vivo. Exp Cell Res 297:108–117
9. Li HZ, Li CW, Li CY, Zhang BF, Li LT, Li JM, Zheng JN, Chang JW (2013) Isolation and identification of renal cell carcinoma-derived peptides associated with GP96. Technol Cancer Res Treat 12:285–293
10. Basu S, Binder RJ, Suto R, Anderson KM, Srivastava PK (2000) Necrotic but not apoptotic cell death releases heat shock proteins, which deliver a partial maturation signal to dendritic cells and activate the NF-kappa B pathway. Int Immunol 12:1539–1546
11. Hunter MC, O'Hagan KL, Kenyon A, Dhanani KC, Prinsloo E, Edkins AL (2014) Hsp90 binds directly to fibronectin (FN) and inhibition reduces the extracellular fibronectin matrix in breast cancer cells. PLoS One 9:e86842
12. Ferrarini M, Heltai S, Zocchi MR, Rugarli C (1992) Unusual expression and localization of heat-shock proteins in human tumor cells. Int J Cancer 51:613–619
13. Srivastava PK, DeLeo AB, Old LJ (1986) Tumor rejection antigens of chemically induced sarcomas of inbred mice. Proc Natl Acad Sci U S A 83:3407–3411
14. Udono H, Srivastava PK (1993) Heat shock protein 70-associated peptides elicit specific cancer immunity. J Exp Med 178:1391–1396
15. Udono H, Srivastava PK (1994) Comparison of tumor-specific immunogenicities of stress-induced proteins gp96, hsp90, and hsp70. J Immunol 152:5398–5403
16. Basu S, Srivastava PK (1999) Calreticulin, a peptide-binding chaperone of the endoplasmic reticulum, elicits tumor- and peptide-specific immunity. J Exp Med 189:797–802
17. Wang XY, Kazim L, Repasky EA, Subjeck JR (2001) Characterization of heat shock protein 110 and glucose-regulated protein 170 as cancer vaccines and the effect of fever-range hyperthermia on vaccine activity. J Immunol 166:490–497
18. Navaratnam M, Deshpande MS, Hariharan MJ, Zatechka DS Jr, Srikumaran S (2001) Heat shock protein-peptide complexes elicit cytotoxic T-lymphocyte and antibody responses specific for bovine herpesvirus 1. Vaccine 19:1425–1434
19. Gong X, Gai W, Xu J, Zhou W, Tien P (2009) Glycoprotein 96-mediated presentation of human immunodeficiency virus type 1 (HIV-1)-specific human leukocyte antigen class I-restricted peptide and humoral immune responses to HIV-1 p24. Clin Vaccine Immunol 16:1595–1600

20. Pawaria S, Binder RJ (2011) CD91-dependent programming of T-helper cell responses following heat shock protein immunization. Nat Commun 2:521. https://doi.org/10.1038/ncomms1524
21. Chandawarkar RY, Wagh MS, Srivastava PK (1999) The dual nature of specific immunological activity of tumor-derived gp96 preparations. J Exp Med 189:1437–1442
22. Chandawarkar RY, Wagh MS, Kovalchin JT, Srivastava P (2004) Immune modulation with high-dose heat-shock protein gp96: therapy of murine autoimmune diabetes and encephalomyelitis. Int Immunol 16:615–624
23. Li X, Liu Z, Yan X, Zhang X, Li Y, Zhao B, Wang S, Zhou X, Gao GF, Meng S (2013) Induction of regulatory T cells by high-dose gp96 suppresses murine liver immune hyperactivation. PLoS One 8:e68997
24. Cohen IR (2014) Activation of benign autoimmunity as both tumor and autoimmune disease immunotherapy: a comprehensive review. J Autoimmun 54:112–117
25. Wick G, Jakic B, Buszko M, Wick MC, Grundtman C (2014) The role of heat shock proteins in atherosclerosis. Nat Rev Cardiol 11:516–529
26. Ueda G, Tamura Y, Hirai I, Kamiguchi K, Ichimiya S, Torigoe T, Hiratsuka H, Sunakawa H, Sato N (2004) Tumor-derived heat shock protein 70-pulsed dendritic cells elicit tumor-specific cytotoxic T lymphocytes (CTLs) and tumor immunity. Cancer Sci 95:248–253
27. Ishii T, Udono H, Yamano T, Ohta H, Uenaka A, Ono T, Hizuta A, Tanaka N, Srivastava PK, Nakayama E (1999) Isolation of MHC class I-restricted tumor antigen peptide and its precursors associated with heat shock proteins hsp70, hsp90, and gp96. J Immunol 162:1303–1309
28. Rivoltini L, Castelli C, Carrabba M, Mazzaferro V, Pilla L, Huber V, Coppa J, Gallino G, Scheibenbogen C, Squarcina P, Cova A, Camerini R, Lewis JJ, Srivastava PK, Parmiani G (2003) Human tumor-derived heat shock protein 96 mediates in vitro activation and in vivo expansion of melanoma- and colon carcinoma-specific T cells. J Immunol 171:3467–3474
29. Suto R, Srivastava PK (1995) A mechanism for the specific immunogenicity of heat shock protein-chaperoned peptides. Science 269:1585–1588
30. Nieland TJ, Tan MC, Monne-van Muijen M, Koning F, Kruisbeek AM, van Bleek GM (1996) Isolation of an immunodominant viral peptide that is endogenously bound to the stress protein GP96/GRP94. Proc Natl Acad Sci U S A 93:6135–6139
31. Blachere NE, Li Z, Chandawarkar RY, Suto R, Jaikaria NS, Basu S, Udono H, Srivastava PK (1997) Heat shock protein-peptide complexes, reconstituted in vitro, elicit peptide-specific cytotoxic T lymphocyte response and tumor immunity. J Exp Med 186:1315–1322
32. Heikema A, Agsteribbe E, Wilschut J, Huckriede A (1997) Generation of heat shock protein-based vaccines by intracellular loading of gp96 with antigenic peptides. Immunol Lett 57:69–74
33. Zügel U, Sponaas AM, Neckermann J, Schoel B, Kaufmann SH (2001) gp96-peptide vaccination of mice against intracellular bacteria. Infect Immun 69:4164–4167
34. Meng SD, Gao T, Gao GF, Tien P (2001) HBV-specific peptide associated with heat-shock protein gp96. Lancet 357:528–529
35. Arnold D, Faath S, Rammensee H, Schild H (1995) Cross-priming of minor histocompatibility antigen-specific cytotoxic T cells upon immunization with the heat shock protein gp96. J Exp Med 182:885–889
36. Arnold D, Wahl C, Faath S, Rammensee HG, Schild H (1997) Influences of transporter associated with antigen processing (TAP) on the repertoire of peptides associated with the endoplasmic reticulum-resident stress protein gp96. J Exp Med 186:461–466
37. Breloer M, Marti T, Fleischer B, von Bonin A (1998) Isolation of processed, H-2Kb-binding ovalbumin-derived peptides associated with the stress proteins HSP70 and gp96. Eur J Immunol 28:1016–1021
38. Binder RJ, Kelly JB 3rd, Vatner RE, Srivastava PK (2007) Specific immunogenicity of heat shock protein gp96 derives from chaperoned antigenic peptides and not from contaminating proteins. J Immunol 179:7254–7261

39. Zhu X, Zhao X, Burkholder WF, Gragerov A, Ogata CM, Gottesman ME, Hendrickson WA (1996) Structural analysis of substrate binding by the molecular chaperone DnaK. Science 272:1606–1614
40. Dollins DE, Warren JJ, Immormino RM, Gewirth DT (2007) Structures of GRP94-nucleotide complexes reveal mechanistic differences between the hsp90 chaperones. Mol Cell 28:41–56
41. Chouquet A, Païdassi H, Ling WL, Frachet P, Houen G, Arlaud GJ, Gaboriaud C (2011) X-ray structure of the human calreticulin globular domain reveals a peptide-binding area and suggests a multi-molecular mechanism. PLoS One 6:e17886
42. Binder RJ (2009) Hsp receptors: the cases of identity and mistaken identity. Curr Opin Mol Ther 11(1):62–71
43. Binder RJ, Han DK, Srivastava PK (2000) CD91: a receptor for heat shock protein gp96. Nat Immunol 1:151–155
44. Basu S, Binder RJ, Ramalingam T, Srivastava PK (2001) CD91 is a common receptor for heat shock proteins gp96, hsp90, hsp70, and calreticulin. Immunity 14:303–313
45. Matsutake T, Sawamura T, Srivastava PK (2010) High efficiency CD91- and LOX-1-mediated re-presentation of gp96-chaperoned peptides by MHC II molecules. Cancer Immun 10:7
46. Tobian AA, Canaday DH, Boom WH, Harding CV (2004) Bacterial heat shock proteins promote CD91-dependent class I MHC cross-presentation of chaperoned peptide to CD8+ T cells by cytosolic mechanisms in dendritic cells versus vacuolar mechanisms in macrophages. J Immunol 172:5277–5286
47. Binder RJ, Srivastava PK (2004) Essential role of CD91 in re-presentation of gp96-chaperoned peptides. Proc Natl Acad Sci U S A 101:6128–6133
48. Tobian AA, Canaday DH, Harding CV (2004) Bacterial heat shock proteins enhance class II MHC antigen processing and presentation of chaperoned peptides to CD4+ T cells. J Immunol 173:5130–5137
49. Leone P, Berardi S, Frassanito MA, Ria R, De Re V, Cicco S, Battaglia S, Ditonno P, Dammacco F, Vacca A, Racanelli V (2015) Dendritic cells accumulate in the bone marrow of myeloma patients where they protect tumor plasma cells from CD8+ T-cell killing. Blood 126:1443–1451
50. Salimu J, Spary LK, Al-Taei S, Clayton A, Mason MD, Staffurth J, Tabi Z (2015) Cross-presentation of the oncofetal tumor antigen 5T4 from irradiated prostate cancer cells—a key role for heat-shock protein 70 and receptor CD91. Cancer Immunol Res 3:678–688
51. Wan T, Zhou X, Chen G, An H, Chen T, Zhang W, Liu S, Jiang Y, Yang F, Wu Y, Cao X (2004) Novel heat shock protein Hsp70L1 activates dendritic cells and acts as a Th1 polarizing adjuvant. Blood 103:1747–1754
52. Messmer MN, Pasmowitz J, Kropp LE, Watkins SC, Binder RJ (2013) Identification of the cellular sentinels for native immunogenic heat shock proteins in vivo. J Immunol 191:4456–4465
53. De Filippo A, Binder RJ, Camisaschi C, Beretta V, Arienti F, Villa A, Della Mina P, Parmiani G, Rivoltini L, Castelli C (2008) Human plasmacytoid dendritic cells interact with gp96 via CD91 and regulate inflammatory responses. J Immunol 181:6525–6535
54. Staudt ND, Jo M, Hu J, Bristow JM, Pizzo DP, Gaultier A, VandenBerg SR, Gonias SL (2013) Myeloid cell receptor LRP1/CD91 regulates monocyte recruitment and angiogenesis in tumors. Cancer Res 73:3902–3912
55. Becker L, Liu NC, Averill MM, Yuan W, Pamir N, Peng Y, Irwin AD, Fu X, Bornfeldt KE, Heinecke JW (2012) Unique proteomic signatures distinguish macrophages and dendritic cells. PLoS One 7:e33297
56. Vénéreau E, Ceriotti C, Bianchi ME (2015) DAMPs from cell death to new life. Front Immunol 6:422
57. Kawashima A, Tanigawa K, Akama T, Wu H, Sue M, Yoshihara A, Ishido Y, Kobiyama K, Takeshita F, Ishii KJ, Hirano H, Kimura H, Sakai T, Ishii N, Suzuki K (2011) Fragments of genomic DNA released by injured cells activate innate immunity and suppress endocrine function in the thyroid. Endocrinology 152:1702–1712

58. Tanaka T, Okuya K, Kutomi G, Takaya A, Kajiwara T, Kanaseki T, Tsukahara T, Hirohashi Y, Torigoe T, Hirata K, Okamoto Y, Sato N, Tamura Y (2015) Heat shock protein 90 targets a chaperoned peptide to the static early endosome for efficient cross-presentation by human dendritic cells. Cancer Sci 106:18–24
59. Srivastava PK, Callahan MK, Mauri MM (2009) Treating human cancers with heat shock protein-peptide complexes: the road ahead. Expert Opin Biol Ther 9:179–186
60. Zhou YJ, Messmer MN, Binder RJ (2014) Establishment of tumor-associated immunity requires interaction of heat shock proteins with CD91. Cancer Immunol Res 2:217–228
61. De Maio A (2011) Extracellular heat shock proteins, cellular export vesicles, and the stress observation system: a form of communication during injury, infection, and cell damage. It is never known how far a controversial finding will go! Dedicated to Ferruccio Ritossa. Cell Stress Chaperones 16:235–249
62. Huang QQ, Pope RM (2013) The role of glycoprotein 96 in the persistent inflammation of rheumatoid arthritis. Arch Biochem Biophys 530:1–6
63. Martin CA, Carsons SE, Kowalewski R, Bernstein D, Valentino M, Santiago-Schwarz F (2003) Aberrant extracellular and dendritic cell (DC) surface expression of heat shock protein (hsp)70 in the rheumatoid joint: possible mechanisms of hsp/DC-mediated cross-priming. J Immunol 171:5736–5742
64. Janeway CA Jr, Medzhitov R (2002) Innate immune recognition. Annu Rev Immunol 20:197–216
65. Ehrlich P (1906) Collected studies on immunity. John Wiley & Sons, London
66. Bashford E, Murray J, Haaland M (1908) Resistance and susceptibility to inoculated cancer. In: Bashford E (ed) Third scientific report on the investigations of the imperial cancer research fund. Taylor & Francis, London, pp 359–397
67. North RJ, Kirstein DP (1977) T-cell-mediated concomitant immunity to syngeneic tumors. I. Activated macrophages as the expressors of nonspecific immunity to unrelated tumors and bacterial parasites. J Exp Med 145:275–292
68. Shankaran V, Ikeda H, Bruce AT, White JM, Swanson PE, Old LJ, Schreiber RD (2001) IFNgamma and lymphocytes prevent primary tumour development and shape tumour immunogenicity. Nature 410:1107–1111
69. Johnson L, Mercer K, Greenbaum D, Bronson RT, Crowley D, Tuveson DA, Jacks T (2001) Somatic activation of the K-ras oncogene causes early onset lung cancer in mice. Nature 410:1111–1116
70. Gatti RA, Good RA (1971) Occurrence of malignancy in immunodeficiency diseases. A literature review. Cancer 28:89–98
71. Schreiber RD, Old LJ, Smyth MJ (2011) Cancer immunoediting: integrating immunity's roles in cancer suppression and promotion. Science 331:1565–1570
72. Binder RJ, Srivastava PK (2005) Peptides chaperoned by heat-shock proteins are a necessary and sufficient source of antigen in the cross-priming of $CD8^+$ T cells. Nat Immunol 6:593–599
73. Li M, Davey GM, Sutherland RM, Kurts C, Lew AM, Hirst C, Carbone FR, Heath WR (2001) Cell-associated ovalbumin is cross-presented much more efficiently than soluble ovalbumin in vivo. J Immunol 166:6099–6103
74. Schumacher TN, Schreiber RD (2015) Neoantigens in cancer immunotherapy. Science 348:69–74
75. Srivastava PK (2015) Neoepitopes of cancers: looking back, looking ahead. Cancer Immunol Res 3:969–977
76. Delamarre L, Mellman I, Yadav M (2015) Cancer immunotherapy. Neo approaches to cancer vaccines. Science 348:760–761
77. Srivastava PK (1996) Do human cancers express shared protective antigens? or the necessity of remembrance of things past. Semin Immunol 8:295–302
78. Nasti TH, Rudemiller KJ, Cochran JB, Kim HK, Tsuruta Y, Fineberg NS, Athar M, Elmets CA, Timares L (2015) Immunoprevention of chemical carcinogenesis through early recognition of oncogene mutations. J Immunol 194:2683–2695

79. North RJ (1984) The murine antitumor immune response and its therapeutic manipulation. Adv Immunol 35:89–155
80. Norbury CC, Basta S, Donohue KB, Tscharke DC, Princiotta MF, Berglund P, Gibbs J, Bennink JR, Yewdell JW (2004) CD8+ T cell cross-priming via transfer of proteasome substrates. Science 304:1318–1321
81. Wolfers J, Lozier A, Raposo G, Regnault A, Théry C, Masurier C, Flament C, Pouzieux S, Faure F, Tursz T, Angevin E, Amigorena S, Zitvogel L (2001) Tumor-derived exosomes are a source of shared tumor rejection antigens for CTL cross-priming. Nat Med 7:297–303
82. Albert ML, Sauter B, Bhardwaj N (1998) Dendritic cells acquire antigen from apoptotic cells and induce class I-restricted CTLs. Nature 392:86–89
83. Clayton A, Turkes A, Navabi H, Mason MD, Tabi Z (2005) Induction of heat shock proteins in B-cell exosomes. J Cell Sci 118:3631–3638
84. Dolan BP, Gibbs KD Jr, Ostrand-Rosenberg S (2006) Dendritic cells cross-dressed with peptide MHC class I complexes prime CD8+ T cells. J Immunol 177:6018–6024
85. Campana S, De Pasquale C, Carrega P, Ferlazzo G, Bonaccorsi I (2015) Cross-dressing: an alternative mechanism for antigen presentation. Immunol Lett 168:349–354
86. Ashley DM, Faiola B, Nair S, Hale LP, Bigner DD, Gilboa E (1997) Bone marrow-generated dendritic cells pulsed with tumor extracts or tumor RNA induce antitumor immunity against central nervous system tumors. J Exp Med 186:1177–1182
87. Neijssen J, Herberts C, Drijfhout JW, Reits E, Janssen L, Neefjes J (2005) Cross-presentation by intercellular peptide transfer through gap junctions. Nature 434:83–88
88. Zhou YJ, Binder RJ (2014) The heat shock protein-CD91 pathway mediates tumor immunosurveillance. Oncoimmunology 3:e28222
89. Tamura Y, Peng P, Liu K, Daou M, Srivastava PK (1997) Immunotherapy of tumors with autologous tumor-derived heat shock protein preparations. Science 278:117–120
90. Gasser S, Raulet DH (2006) The DNA damage response arouses the immune system. Cancer Res 66:3959–3962
91. Lauren B. Kinner-Bibeau, Abigail L. Sedlacek, Michelle N. Messmer, Simon C. Watkins, Robert J. Binder (2017) HSPs drive dichotomous T-cell immune responses via DNA methylome remodelling in antigen presenting cells. Nature Communications 8:15648
92. Abigail L. Sedlacek, Lauren B. Kinner-Bibeau, Robert J. Binder (2016) Phenotypically distinct helper NK cells are required for gp96-mediated anti-tumor immunity. Scientific Reports 6 (1)

# Chapter 5
# Bridging the Gaps in the Vaccine Development: Avant-Garde Vaccine Approach with Secreted Heat Shock Protein gp96-Ig

**Natasa Strbo**

*Dedicated to my teacher, mentor and friend who was always*
***Ahead of his time***
*Eckhard R Podack*

*(February 1943–October 2015)*

**Abstract** Design of highly pure and safe vaccines in post-genomic era unfortunately includes the inherent lack of immunostimulatory properties of proteins and peptides. Vaccine adjuvants are therefore considered key components in modern vaccinology since they provide the necessary help of enhancing the immune responses. Over the past two decades, Dr. Podack's laboratory has developed an exciting and avant-garde reagent: a heat shock protein-based vaccine, chaperone gp96, that generates effective antitumor and anti-infectious immunity in vivo. State-of-the-art secreted gp96-Ig vaccine provides within one molecule strong adjuvant properties and antigen specificity for cross-priming CD8 T cells and activation of innate immunity. Gp96-peptide complexes were identified as an extremely efficient, femto-molar pathway of MHC I-mediated antigen cross-presentation, generating CD8 CTL responses detectable in the blood, spleen, liver, intestinal and reproductive tract lamina propria, and intraepithelial compartment, respectively. These studies provided the first evidence that cell-based gp96-Ig-secreting vaccines may serve as a potent modality to induce not only systemic but also mucosal immunity. The gp96-Ig vaccine strategy has been utilized in clinical trials for non-small cell lung cancer (NSCLC) patients and as prophylactic SIV vaccine to protect nonhuman primates from mucosal infection upon challenge with SIV, demonstrating the feasibility and benefits of this approach for both safety and efficacy.

**Keywords** Heat shock proteins · gp96 · Vaccine · Cancer · HIV · Immunotherapy

N. Strbo (✉)
Department of Microbiology and Immunology, Miller School of Medicine,
University of Miami, Miami, FL, USA
e-mail: nstrbo@med.miami.edu



R. J. Binder, P. K. Srivastava (eds.), *Heat Shock Proteins in the Immune System*,
https://doi.org/10.1007/978-3-319-69042-1_5

## 5.1 Introduction

Vaccines represent one of the greatest triumphs of modern medicine. Over the past centuries, vaccine development has followed Pasteur's principles of isolation, inactivation, and injection of pathogen organism [1]. This led to the development of bacterial and viral vaccines that are composed of whole killed pathogens, live-attenuated pathogens, or purified components from pathogens (subunit vaccines and recombinant proteins/peptides/DNA vaccines) (Fig. 5.1).

The goal of vaccination is the generation of strong immune response to the administrated antigen able to provide a long-term protection against infection. Type of the immunity that can be induced by vaccination are antibody or T cell-specific immune responses. Most vaccines licensed so far induce antibodies [1, 2]. On the other side, a literature suggests that cytotoxic T cells are important in protection from infectious diseases (HIV, TB) and cancer [3–5].

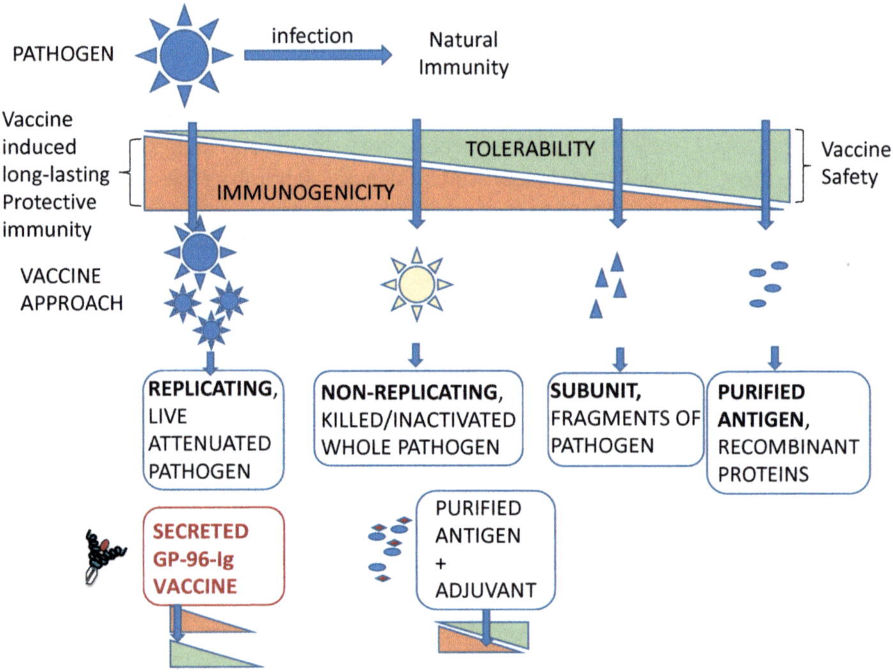


**Fig. 5.1** Balancing immunogenicity and tolerability by different vaccine approaches. The effect of different vaccine approaches (replicating live-attenuated pathogens, non-replicating killed/inactivated whole pathogens, subunit or purified antigens, secreted gp96-Ig and purified antigen with adjuvant) on the vaccine-induced immunogenicity/protective immunity and vaccine tolerability/safety. Improving the anticipated benefits (efficacy) of immunization while decreasing their potential risks (adverse reactions) underpins the development of all new vaccines and is a key factor driving innovative vaccine design such as secreted gp96-Ig vaccine approach

Although most vaccines induce good memory responses, the type of memory induced by different vaccines may be considerably different. Immune responses to natural infection and vaccination suggest that the type and duration of immune memory are largely determined by the magnitude and complexity of innate immune signals that imprint the acquired immune primary responses. In addition, while vaccines containing a limited number of purified antigens generally have improved tolerability/safety profiles compared with live-attenuated and whole-pathogen vaccines, they are also often less immunogenic due to the removal of pathogenic features of the organism (Fig. 5.1) [6]. Improving the anticipated benefits (efficacy) of immunization while decreasing their potential risks (adverse reactions) underpins the development of all new vaccines and is a key factor driving innovative vaccine design such as secreted gp96-Ig vaccine approach [7].

Furthermore, existing vaccines for infectious disease have been developed mostly against pathogens that show no or limited antigenic variation and that can be controlled by neutralizing serum antibodies. In contrast, the conquest of pathogens that display more variable antigens (*HIV*, *M. Tuberculosis*, *P. falciparum*) and require T cell immunity remains elusive. The vaccine principles necessary for the generation of appropriately activated cellular immunity mediated by CD8+ cytotoxic T cells for infectious diseases also apply to therapeutic cancer vaccines. Effective cancer immunotherapy is widely believed to originate with appropriately activated CD8+ cytotoxic T cells (CTL) to tumor antigens displayed on MHC I; however, the vast majority of cancer vaccine approaches in development lead to preferential display of vaccine antigen (either purified or cell based) on MHC II following macrophage-mediated phagocytosis of vaccine cells or protein.

The innovative approach taken by our laboratory relies on secreted gp96-Ig chaperoning infectious or tumor antigenic proteins that are efficiently taken up by activated APCs and cross-presented via MHC I to CD8 CTL, thereby stimulating an avid, antigen-specific, cytotoxic CD8 T cell response (Figs. 5.1 and 5.2). The immune system has evolved to recognize free gp96-peptide complexes and other chaperones and uses chaperoned antigenic peptides to activate both arms of the immune system: innate and adaptive arm (Fig. 5.2). Current knowledge suggests that antigen-presenting cells (APCs), such as dendritic cells, play a key intermediary role between the innate and adaptive responses and are critical in determining the direction of the adaptive immune response [8]. The ideal vaccine, therefore, would initiate an innate immune response capable of directing the adaptive immune response toward efficient activation and removal of the specific pathogen, followed by the development of immune memory. The limited ability of highly purified vaccines to induce protective immunity appears to be related to their failure to induce maturation of APCs. Gp96-Ig-secreted vaccines are capable of initiating both innate immunity (activation of APCs, pro-inflammatory cytokine release, activation of NK cells) and adaptive immune responses (priming, activation, and proliferation of antigen-specific CTLs) that lead to successful clearance of the antigen/pathogen (Fig. 5.2). Gp96-peptide complexes are internalized via receptor-mediated endocytosis (CD91, TLR2/4 and SRA, LOX-1). The peptides enter the MHC pathway of

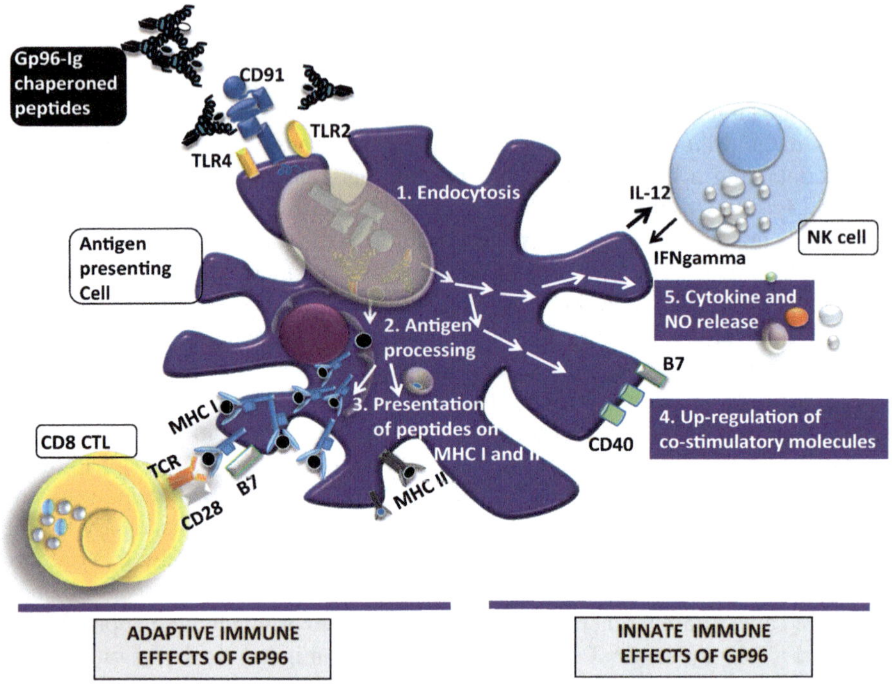


**Fig. 5.2** The effect of gp96-peptide complex on the antigen-presenting cell (APC). Gp96-chaperoned peptides are released from cells only upon necrotic cell death. Gp96-peptide complexes are internalized via receptor-mediated endocytosis (CD91, TLR2/4 and SRA, LOX-1). The peptides enter the MHC pathway of antigen processing and presentation. Ultimately, the peptides chaperoned by gp96 are presented on MHC I and MHC II molecules (adaptive immune effects). Gp96 also trigger signaling receptors to activate NF-kb. APCs mature and upregulate costimulatory molecules and release cytokines, chemokines, and NO (innate immune effects)

antigen processing and presentation. Ultimately, the peptides chaperoned by gp96 are presented on MHC I and MHC II molecules (adaptive effects) (Fig. 5.2). A second fundamental property of gp96 is an inherent adjuvanticity, which is secondary to gp96 interaction with toll-like receptors 2 and 4. Gp96 binding to TLR2 and TLR4 on dendritic cells and macrophages leads to upregulation of costimulatory molecules B7-1, B7-2, CD40, and MHC II and the release of cytokines and chemokines IL-12, IL-1β, TNF-α, RANTES, MCP-1, and nitric oxide [9–13]. These outcomes involve the stimulation of the central signaling molecule NF-κb and its translocation into the nucleus [9]. In vivo, APCs are also stimulated to migrate to regional lymph nodes [14, 15].

Above-described gp96 immune properties have been used successfully in murine models of cancer, in nonhuman primates for SIV prophylaxis, and in clinical trials for the treatment of non-small cell lung cancer patients.

## 5.2 The Visionary Work of Dr. Eckhard Podack and Development of the State-of-the-Art Secreted gp96-Ig Vaccine Approach

During his scientific career, Dr. Eckhard Podack made numerous seminal discoveries that have contributed to our understanding of innate and adaptive immunity [16–36]. In the mid-1980s, he established the mechanism of eukaryotic cell killing by cytotoxic T cells and natural killer cells. This work led to the discovery of a distinct pore-forming protein in cytotoxic T cells [28]. Dr. Podack named this new molecule perforin, which forms transmembrane channels in target cells to induce their cellular lysis. Perforin is a critical component of cytotoxicity to fight viral infection and is involved in tumor and transplant rejection and in the pathogenicity of autoimmune diseases [23, 24]. To translate his discoveries regarding lymphoid killing pathways to therapeutic cancer intervention, in the mid-1990s, he developed vaccine that generates potent CD8 cytotoxic T lymphocyte (CTL) responses [37]. The strategy he employed was based on the process of antigen cross-presentation and the role that one of heat shock chaperons, gp96, plays in this process. He has been able to modify gp96 into a secretory chaperone protein which, because of its various properties, dramatically augments the frequency of antigen-specific CTL. The degree of CTL expansion seen with gp96-Ig vaccination exceeds that seen with any other vaccine approach to date and suggests that this approach using a novel secreted form of heat shock protein gp96 could be used as preventive as well as therapeutic tool to significantly augment antitumor and antiviral immunity [7]. Applying this general concepts across a broad spectrum of diseases, Dr. Podack developed gp96-based cellular vaccines for cancer (non-small cell lung carcinoma (NSCLC), bladder cancer, pancreatic and ovarian carcinoma) [38], SIV and HIV [35], and malaria, and currently, work is in progress to develop gp96-Ig vaccines against leishmania and zika virus (Fig. 5.3). Studies are ongoing in experimental systems in the laboratory and this approach has moved into the clinic, including phase 2 trials for bladder cancer and phase 1b for NSCLC. As cell-based vaccines, the gp96-Ig approach met with skepticism but over time overcame hurdles and was accepted by FDA regulation resulting in the formation of a Biotech Startup company (Heat Biologics) conducting clinical studies.

In this review article, we want to acknowledge Dr. Podack's innovative and visionary work in the gp96 vaccine development arena (Fig. 5.3).

### *5.2.1 Construction of gp96-Ig*

In the Journal of Immunology in November 15, 1999, vol. 163 no. 10 (pg. 5178–5182) [37], Dr. Podack and his team Dr. Yamazaki and Nguyen have reported for the first time that modification of endoplasmic reticulum heat shock protein gp96 to KDEL-deleted gp96-Ig fusion protein resulted in the secretion of gp96 together

**Fig. 5.3** Secreted gp96-Ig vaccine approach and Dr. Eckhard Podack visionary vaccine development work

with bound peptides from transfected tumor cells. To generate the gp96-Ig fusion protein, the KDEL sequence was deleted and replaced with the hinge, CH2 and CH3 domains of murine IgG1 (16–23) (Fig. 5.3); double-stranded gp96 cDNA was prepared from Jurkat DNA (24) and amplified by PCR. The hinge, CH2 and CH3 domains of murine IgG1, was amplified by using murine IgG1 cDNA as a template and mutating the three cysteines of the hinge portion to serines (21, 25). Gp96 was inserted into eukaryotic expression vector, pBCMGSNeo and pBCMGHis (26–29), and transfected into different mouse cell lines, SCLC-2, SCLC-7, B16F10, MC57, LLC NIH3T3, EL4, E.G7, and P815, and human cell lines HEK-293, AD100, MiaPaca, and JEG-3.

Tumor cell lines transfected with secreted form of gp96 were more immunogenic, and secreted gp96-Ig was responsible for decreased tumorigenicity. In this initial study, it was confirmed that immunization with tumor cell-secreting gp96-Ig generates efficient tumor-rejecting CD8 cytotoxic T cells (CTLs) without requirement for CD4 or macrophage help [25, 37]. Later on it was also confirmed that the ability of the vaccine to stimulate CTL expansion is significantly inhibited in the presence of an established tumor [39, 41]; however, this study also demonstrated that more frequent vaccinations were sufficient to retard growth of the established tumors.

### *5.2.2 Principles of Secreted gp96-Ig Vaccine Approach*

Heat shock proteins/chaperones and their chaperoned peptides are released from cells primarily upon necrotic cell death that may be caused by infection, trauma, nutrient deprivation, extreme cell stress, or by tumor necrosis [9]. Gp96 has evolved to serve as a molecular warning signal for necrotic cell death. The unique localization of gp96 to the ER, and local exposure to all peptides destined for presentation on MHC I, likely influenced the acquisition of both the dual antigen-delivery and antigen-presenting cell adjuvant properties of gp96. These properties make gp96 one of the few endogenous signals that can both **activate and deliver antigen to APCs** (Fig. 5.2). Further, the specific transfer of gp96 chaperoned antigens by antigen-presenting cells to MHC I through the cross-presentation pathway endow gp96 with truly unique characteristics as a basis for a CD8+ T cell-specific vaccine protein (Fig. 5.2). To take advantage of this unique adjuvant effect and ability to transport relevant peptides, we set up a model system that imitates necrotic cell death with regard to the release of HSP (Fig. 5.4a and b). This system allowed us to analyze the immunological effects of HSP in vivo independent of infectious agents and cell death.

The bulk of proteins in the ER are destined for secretion or for insertion into the plasma membrane or membranes of other cellular organelles. Proteins residing permanently in the ER, to which group gp96 belongs, are retained there by the KDEL retention signal usually located at the C-terminus of the protein (Fig. 5.4a). Replacing the KDEL sequence of gp96 with the hinge, CH2 and CH3 domain of murine IgG1, an Ig isotype inefficient in Fc receptor binding, and transfection of the cDNA into cells resulted in the secretion of gp96-Ig into the culture supernatant where it was quantitated by Western blotting and ELISA (Figs. 5.3 and 5.4b). Western blotting with a monoclonal antibody specific for gp96 confirmed the identity of the fusion protein, predicted molecular mass of 120 kDa (Fig. 5.4b) [37, 40]. In model systems in mice, we have shown that gp96-Ig-transfected, antigen-expressing cells secrete gp96-Ig in vivo and stimulate cognate systemic cellular CD8 CTL immune responses. Generated antigen specific CD8 memory responses are independent of CD4 help and CD40L and can be established in the absence of lymph nodes [25, 33, 41, 42].

As shown in Figs. 5.4 and 5.5, gp96 is a dimer, able to provide a protected cavity for protein folding and peptide binding [43, 44]. Removal of peptides chaperoned by HSPs abrogates the immunogenicity of HSP preparations [45, 46]. The peptide-binding properties of gp96 have been investigated by several groups and found to be both unusually promiscuous and highly stable [47–49]. There is disagreement about the location of the peptide-binding domain which has been located to both the C-terminal portion of the protein near the dimerization domain [50, 51] and to the N-terminal domain [52–54]. In contrast to MHC I, gp96 is capable of binding peptides with variable length and composition. In addition to binding potentially antigenic peptides, gp96 has a protein-binding domain which is involved in chaperoning the folding of newly synthesized proteins: IgGs, some integrins, and all of toll-like receptors [55–57].

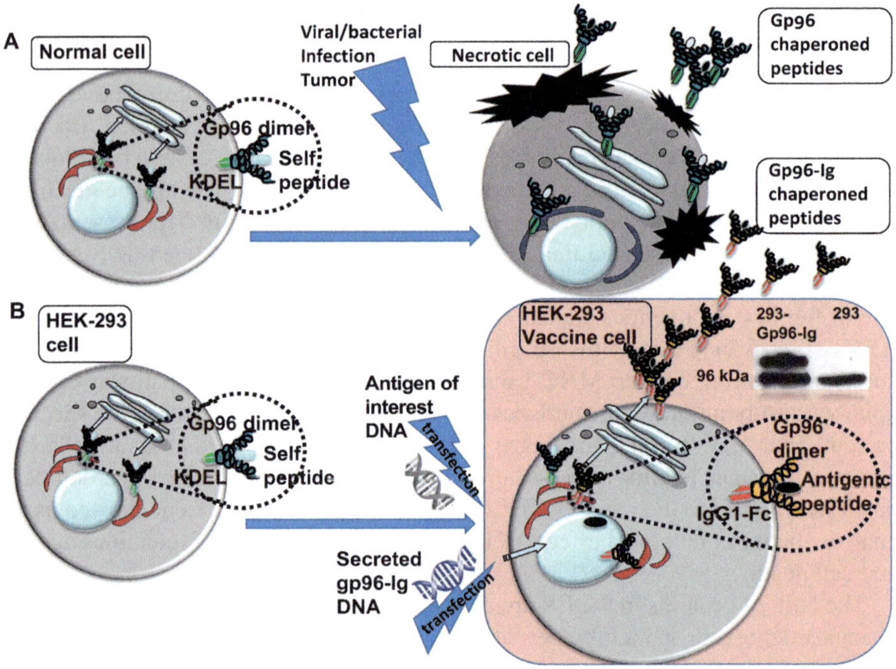


**Fig. 5.4** Secreted gp96-Ig. Gp96 is a naturally occurring protein that stays within all normal and tumor/infected cells (**a**, **b**). The reason gp96 cannot leave living cells is because it contains a retention signal; KDEL Gp96 containing KDEL sequence is retained in ER after sorting in the Golgi apparatus by retrograde transport (**a**, **b**). Only necrotic cells can release gp96 (**a**). We developed genetically modified gp96 by replacing the retention signal with a secretion signal, IgG1-Fc (**b**). Gp96-Ig-chaperoned peptides are exported together with other secreted proteins and can be detected in the cell supernatant by ELISA and Western blotting. Western blotting on whole cell lysate with a monoclonal antibody specific for gp96 confirmed the identity of the fusion protein, predicted molecular mass of 120 kDa and endogenous gp96 of 96 kDa (**b**). Cells that secret gp96-Ig generate a very powerful immune response against its target

The properties and location of gp96 in the ER place it into a strategic position to come into contact with virtually all proteins and peptides that are present in a cell and are used for MHC I loading. If the cell is infected by viruses or other intracellular parasites, gp96 also serves as chaperone for viral and other pathogen proteins [58]. Similarly, gp96 serves as chaperone for tumor-associated antigens expressed by tumor cells [59]. We took advantage of the unique peptide-binding properties of gp96 to develop two different secreted gp96 vaccine strategies (Fig. 5.5)**.** First one is used for the development of antitumor vaccine and relays on the transfection of different tumor cell lines. Second strategy, development of anti-infectious disease vaccine, deploys 293 cells as a vaccine cells. In the antitumor vaccine approach, transfected tumor cells indeed secreted gp96 and, when transplanted into syngeneic

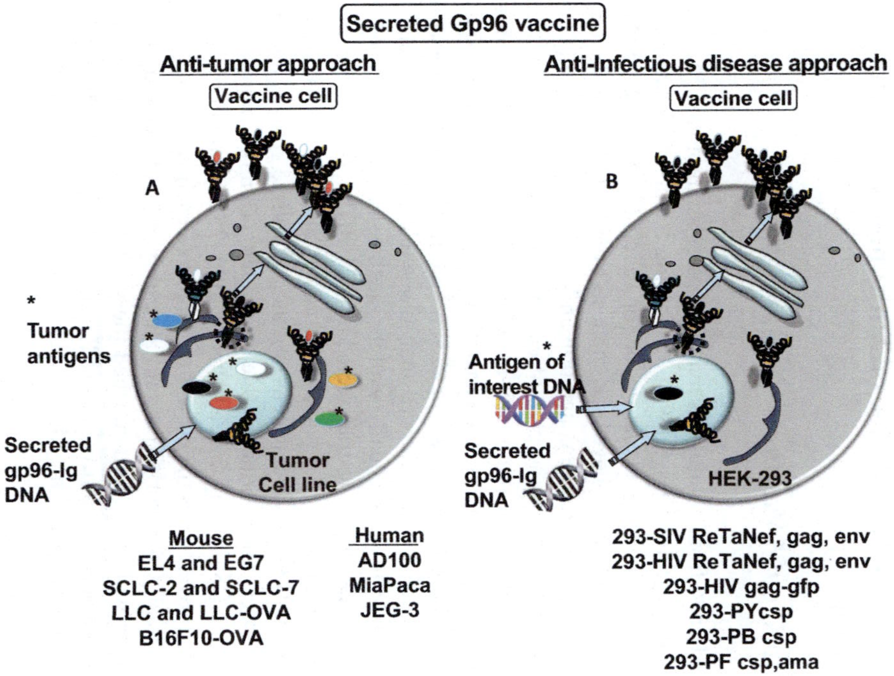


**Fig. 5.5** Two gp96-Ig vaccine approaches: (**a**) antitumor vaccine approach and (**b**) anti-infectious disease vaccine approach. (**a**) Gp96-Ig was transfected into different mouse tumor cell lines, SCLC-2, SCLC-7, B16F10, MC57, LLC, EL4, and E.G7, and human tumor cell lines, AD100, MiaPaca, and JEG-3. (**b**) HEK-293 cells were transfected with secreted gp96-Ig construct and different combinations of infectious antigens: SIV/HIV retanef, gag, env, *Plasmodium falciparum*, *P. berghei*, *P. yoelii* CSP, and AMA1

mice, were rejected by the immune system. The untransfected parental tumor cells, in contrast, grew and killed the mice. Immune rejection was dependent on CD8 but independent of CD4 cells and CD40 ligand [25]. In anti-infectious disease approach, by transfecting 293 cells with gp96-lg and combination of plasmids encoding different infectious antigens (SIV/HIV retanef, gag, gp120, plasmodium antigens), secreted gp96$^{\mathbf{SIV/HIV}}$lg induces a SIV-/HIV-specific CD8 CTL response [40].

## 5.3 The State of the Knowledge of Antitumor Vaccination with gp96-Ig

### *5.3.1 Allogeneic gp96 Vaccine*

The use of purified gp96 from autologous tumor biopsies as autologous tumor vaccine given as a bolus injection has shown encouraging results, as discussed elsewhere [7]. However, the need of viable tumor for vaccine generation has been a limitation, and the clinical results have been modest. In order to bypass the need of viable autologous tumor for vaccine generation and to make a vaccine that can be applied to all patients with the same type of cancer, our group developed a novel gp96-Ig-secreting and allogeneic tumor cell-based vaccine [37]. As described previously, this vaccine strategy stimulates the generation of potent, polyepitope-specific, multi-cytokine-secreting CTL responses against all antigenic tumor epitopes present in the tumor vaccine cell. We showed that allogeneic tumor cells transfected with and secreting gp96-Ig work equally well as vaccine cells as autologous cells. Using irradiated allogeneic gp96-Ig-transfected tumor cells as vaccine, the vaccine cells are still alive but replication is incompetent and will survive for several days in the vaccinated patient. The live vaccine cells continue to secrete gp96-Ig and stimulate CD8 CTL responses. The continuous release of gp96 is a more appropriate and stronger stimulus for CD8 priming than a bolus of gp96 [25]. In addition, the use of allogeneic tumor cells overcomes the need for tumor from each patient and makes a universal vaccine. Finally, the work from our laboratory [41, 39] showing that frequent immunizations with gp96-Ig dampened the growth of certain experimental tumors suggested that such a protocol restored antitumor immunity and overcame tumor-induced immunosuppression. Allogeneic tumor cell-based vaccine strategy is summarized on Fig. 5.6.

The use of allogeneic tumor cells as a source of tumor antigens chaperoned by secreted gp96-Ig and cross-presented via MHC I to patient CD8 T cells is based on the hypothesis that allogeneic tumors have an overlapping repertoire of similar tumor antigens analogous to allogeneic melanomas and allogeneic small cell lung cancers [37, 60–62]. Gene array analyses of many tumor types, including NSCLC tumors [63], support the notion of sharing tumor-associated antigens.

We have developed **a gp96-Ig and allogeneic tumor cell-base vaccine strategy** that:
Stimulates

1. Generation of potent, polyepitope specific, multi-cytokine secreting, CD8 cytotoxic T lymphocytes (CTL) against all antigenic tumor epitopes present in the tumor-vaccine cell.

2. Mediates efficient tumor antigen MHC I cross presentation (at femto-molar antigen concentration) via vaccine-gp96-Ig activated patient dendritic cells (DC) and NK cells and generates potent CD8 CTL response

3. With this strategy, specific knowledge of tumor associated antigens is not required

4. Stimulating the generation of potent allogeneic and tumor-polyepitope specific, multi-cytokine secreting CD8 CTL against any or all antigenic tumor epitopes present in the umor-vaccine cell is a **novel approach to tumor vaccine therapy**

**Fig. 5.6** Principles of gp96-Ig allogeneic tumor cell-base vaccine. Four major effects of gp96-Ig allogeneic tumor cell-based vaccine

Our group conducted a phase 1 trial in stage IIIB/IV non-small cell lung cancer (NSCLC) to evaluate the safety and feasibility of this approach reviewed in Strbo et al. [7]. We also evaluated for the first time in patients the method of frequent vaccination. Patients were vaccinated with a gp96-Ig-secreting allogeneic NSCLC line (AD100-gp96-Ig), irrespective of their HLA type. Although the study is limited in lacking a control arm and in having been closed prematurely by the institution for reasons entirely unrelated to the study, it offers extremely interesting insights into the effects of therapeutic vaccine immunotherapy. Our method of preparing vaccines from established allogeneic tumor cell lines by transfection with gp96-Ig provides a relatively simple and inexpensive way to conduct tumor-vaccine immunotherapy. Off-the-shelf allogeneic vaccines are of great advantage as therapeutic option. Furthermore, in this single institution study of 18 patients, the tumor cell-based gp96-Ig-secreting AD100-gp96-Ig vaccine was found to have an acceptable safety profile, achieving a significant disease control rate in a heavily pretreated population and a substantial CD8 CTL response. Our data indicates that CTL responses are required to obtain a clinical benefit, but that in the majority of patients, the tumor burden was too extensive to achieve complete stop or reverse tumor progression with the gp96-Ig-induced CTL response. Several groups have shown a correlation with tumor burden and elevated level of regulatory T cells (Treg) in circulation, which in turn act to suppress immunity [64, 65]. It is believed now that patients with minimal tumor burden are the best candidates for immunotherapy [66]. Hence, further evaluation in a phase 2 trial, both in NSCLC and other tumor types, in the patients with minimal tumor burden is warranted.

### 5.3.2 Combined Therapeutic Approach

T cells infiltrated in tumor tissue are often capable of recognizing tumor-associated antigen, but they coexist with their target, tumor cells, without significant antitumor activity. However, when isolated from tumor tissue, those tumor-infiltrated T cells could kill the tumor cells efficiently in vitro [67]. These studies indicate that tumors establish a stern environment for antitumor immune cells, cells that can be active effector cells otherwise.

Immunosuppression in the tumor microenvironment involves CTLA-4 and PD-immunosuppressive signal, anti-inflammatory cytokines (IL-10, TGF-β), enzymes (indoleamine-2,3-dioxygenase), and professional immunoregulatory cells [regulatory T cells, myeloid-derived suppressor cells (MDSCs)] [68, 69]. The identification of immunosuppressive mechanisms in tumors pointed out molecular targets to restore the antitumor immune response. Thus, these negative immunoregulatory mechanisms, so-called immune checkpoints, became a focus in drug discovery. The effort resulted in FDA approval of anti-CTLA-4 and anti- PD-1 antibodies for cancer treatment. This achievement finally convinced people that immunotherapy of cancer is realistic, and it further encouraged the development of inhibitors of other immune checkpoint molecules [70–72].

While the current results observed in patients treated with antitumor vaccines are still deceiving, it looks probable that the combination of tumor vaccines with new maneuvers to eliminate tumor-induced immunosuppression [73, 74] will soon lead to a new form of cancer management.

We think that the success of complete tumor rejection is in combined therapeutic approach: combining the *most potent multi-epitope-specific CD8 CTL vaccine with the most potent reversal of tumor-induced immune suppression*. In this combination, multi-epitope-specific CD8 CTLs are generated by gp96-Ig-secreting tumor cells which now can perform their task of killing tumor cells without interference by tumor-induced suppression signals. Extracellular adenosine has been known as an inhibitor of immune functions [75]. When cells are deprived of nutrients or oxygen, insufficient ATP biosynthesis tends to lower the ATP/adenosine ratio. Indeed, tissue hypoxia strongly represses proliferation of activated T cells [76]. Interestingly, extracellular adenosine is known to accumulate under hypoxic conditions. Currently, we are investigating the novel approach of using anti-hypoxia in combination with immunotherapies. Most recently we showed that breathing 60% oxygen, which is a standard medical procedure, can overcome tumor hypoxia [22, 77]. We are the first to show the impressive improvement of immune responses by hyperoxia with 60% oxygen. Moreover, the dramatic effectiveness of gp96-Ig vaccination in combination with 60% oxygen for the treatment of large established tumor burdens is highly innovative, and we believe the combination therapy with gp96-Ig may foretell a breakthrough with significant impact on the cancer field with benefits for patients.

## 5.4 The State of Knowledge of Anti-Infectious Vaccination with gp96-Ig

### *5.4.1 Secreted SIV/HIV gp96-Ig Vaccine*

The most successful vaccines have been against diseases where the causal pathogen does not have major anti-immune defense mechanisms. Many pathogens, including hepatitis C virus (HCV) and human immunodeficiency viruses (HIV), *M. tuberculosis* (TB), and *Plasmodium falciparum* have evolved complex immune evasion strategies and require a high level of effector T cell activation for their eradication. So far, these pathogens have proved intractable to existing vaccination strategies. Heat shock proteins possess significant properties that support their inclusion and testing in the next generation of infectious disease vaccines.

The use of viruses and bacteria or of viral vectors or attenuated viruses for induction and analysis of the immune response relies to a large extent on the ability of viral or bacterial components to activate the immune system (e.g., via pattern recognition receptors). Live-attenuated vaccines against small pox and yellow fever elicit brisk, polyepitope-specific, polyfunctional CD8+ T cell responses that contribute to protection [78, 79]. Similarly, in live-attenuated SHIV-immunized macaques, polyfunctional T cell responses are associated with a better control of challenge virus replication [80, 81]. The cell-based secreted gp96-Ig vaccines, by prolonged in vivo secretion of immunogenic gp96-Ig peptide complexes, resemble viral replication and contribute to the cytotoxic response by providing immune stimuli comparable to attenuated viruses. Although non-viral and non-bacterial in nature, gp96-mediated CD8 CTL responses bear the hall marks of memory responses characteristically seen after viral or bacterial infections. We attribute this observation to the adjuvanticity of gp96 which is specifically directed toward cross-priming, cytotoxic CD8 CTL responses [82].

Our vaccine, denoted as 293-gp96$^{SIV}$Ig, was made by transfecting 293 cells with gp96-lg and plasmids encoding SIV-retanef, SIV-gag, and SIV-gp120 (retanef is a fusion protein of rev nef and tat) provided by Drs. Franchini, Felber, and Pavlakis (NIH) in collaboration. The significance of the cell (HEK-293 cells, not containing T antigen) is that it acts as a "pump" continuously secreting gp96-Ig over 3–4 days and activating immune responses until the cells are rejected by allo- or anti-SIV responses. In effect, continuous secretion of gp96-Ig provides a continuous stimulus quite similar to replicating attenuated viruses, which provide excellent protection when used as vaccines. Thus, when cells containing SIV antigens and secreting gp96$^{SIV}$lg are injected into recipients, secreted gp96$^{SIV}$lg induces a SIV-specific CD8 CTL response. As described above, gp96 is a potent Th1 adjuvant by activating APCs and NK cells (Fig. 5.2). In addition gp96-Ig chaperones client peptides derived from SIV antigens (Fig. 5.5). Gp96-Ig is endocytosed by CD91 on the activated APCs, and the client peptides are cross-presented by MHC I, priming antigen-specific CD8 T cells. Since adjuvant and antigenic peptide (epitope) are part of the same molecular complex, cross-presentation and priming of antigen-specific CD8 T cells are extraordinarily efficient requiring only femto-molar ($10^{-15}$ M) antigenic peptide [25]. Polyepitope specificity is achieved because gp96-Ig carries all client peptides generated from the transfected SIV antigens by the proteasome and translocated by TAP into the ER of the host DC. The client peptides are further trimmed and selected for MHC I presentation by the host DC. Thus, any T cell epitope present in the transfected SIV antigen will be cross-presented by the host MHC I and primes corresponding antigen-specific CD8 T cells. This principle provides the largest degree of polyepitope specificity possible to be presented by any MHC I type. The strong adjuvant and Th1 activity of gp96-Ig provide for multi-cytokine CTL responses [25, 34, 40]. Importantly, self-peptides do not generate CD8 CTL responses due to normal tolerance mechanisms. We have not observed any signs of autoimmunity in mice, macaques, or humans (in vaccine trials for lung cancer) in any of our gp96-lg-based vaccine studies.

#### 5.4.1.1 Gp96$^{SIV}$Ig as a Novel Adjuvant for Antibody Production

Gp96-Ig-chaperoned peptides are cross-presented primarily by MHC I. Accordingly, gp96-Ig-based immunization generates powerful CD8 CTL responses but little antibody [35, 40, 83]. Since gp96-Ig is a potent adjuvant for DC activation, addition of protein antigen, gp120, resulted in uptake by DC via classical endo- or pinocytosis and presentation of processed antigen by MHC II to generate help for B cell antibody production [35]. The gp96-Ig-induced Th1 environment induced isotype switching of B cells to generate IgG1 and IgG3 antibodies that bind to Fc receptors on macrophages and NK cells and activate complement and may contribute to SIV virus neutralization.

We used transcriptional profiling of host responses to vaccination and subsequent repeated SIV challenges to further understand the protective effects of a novel secreted gp96-Ig vaccination strategy [84]. Focusing on postchallenge comparisons, in particular for protected animals, we identified a host response signature of protection comprised of strong interferon signaling after the first challenge, which then largely abated after further challenges [84]. We also identified a host response signature, comprised of early macrophage-mediated inflammatory responses, in animals with undetectable viral loads 5 days after the first challenge but with unusually high viral titers after subsequent challenges. Statistical analysis showed that prime-boost vaccination significantly lowered the probability of infection in a time-consistent manner throughout several challenges. Given that humoral responses in the prime-boost group were highly significant prechallenge correlates of protection, the strong innate signaling after the first challenge suggests that interferon signaling may enhance vaccine-induced antibody responses and is an important contributor to protection from infection during repeated low-dose exposure to SIV [84].

#### 5.4.1.2 Protection from Mucosal SIVmac251 Infection Requires Strong Mucosal CTL and Antibody Responses

The ultimate goal of the HIV research community is the development of protective vaccine. Only one HIV vaccine trial in humans (RV144), to date, has demonstrated partial vaccine efficacy [85], and vaccine protection was achieved in the absence of a strong neutralizing antibody response. With the novel combination of cell-secreted **gp96$^{SIV}$Ig** and **gp120 protein** as immunogen, we have achieved significant reduction (73%) in the risk for SIV acquisition in nonhuman primates [35]. Protection from mucosal SIVmac251 infection was associated with strong mucosal CTL and non-neutralizing antibody responses. The antibody isotype suggested TH1 polarization. Our control vaccines that generated either SIV-specific CTL or SIV-specific antibodies had no protective effect suggesting that specific CTL and antibodies are required for protection.

Using the novel principle for generating SIV-specific Th1 CTL and Th1 antibodies described above, we have achieved for the first time significant protection (73%) against infection by up to 7 weekly rectal challenges with highly pathogenic SIVmac251 (Fig. 5.5) [35]. Unlike SIVmac239 which is a cloned virus, SIVmac251 is not cloned and has considerable sequence diversity. In our study, the combination of gp96$^{SIV}$Ig with gp120-protein was required for protection. CTL alone or antibody alone did not provide protection [35]. 73% immunization efficiency is an encouraging starting point for further development of the immunization strategy and to understand which immune responses are required for better protection from SIVmac251. We hypothesize that mucosal antibody in the mucus of rectum and vagina will trap the virus in the mucus antibody network, preventing contact with mucosal cells and hence preventing infection. The viruses that penetrate the barrier reaching and infecting cells require CTL or NK or other cytotoxic responses to eliminate infected cells to prevent viral replication and spreading.

#### 5.4.1.3 DNA-Based Vaccination in Association with Electroporation (EP)

An important component of our gp96$^{HIV/SIV}$-Ig vaccine strategy is the use of live but irradiated cells as vaccines that after injection act as a "pump" continuously secreting gp96-Ig for several days (average 3–5), thereby activating immune responses in the same fashion as attenuated viral vaccines do. Recently, we utilize an attractive mode of vaccine delivery, DNA-based vaccination in association with electroporation (EP). Muscle cells are co-transfected with gp96-lg, HIV/SIV antigens by in vivo *electroporation.* Upon transfection the muscle cells continuously secrete gp96-Ig chaperoning HIV/SIV peptides that are MHC I cross-presented by gp96-activated DC to CD8 cells generating potent HIV/SIV-specific CTL. We have shown previously that cell-based gp96-lg-mediated cross-priming and clonal expansion of CD8 T cells by DC are ~20 million-fold more efficient compared to cross-priming by protein antigen alone [25]. This efficiency is achieved by the fact that gp96 unites in one molecule strong adjuvant activity and antigen specificity by chaperoning antigenic (SIV/HIV) peptides.

### 5.4.2 Secreted gp96-Ig Malaria Vaccines

Malaria infects nearly 250 million people annually and causes almost one million deaths. An effective vaccine against malaria would be a valuable public health tool, complementing anti-malaria drugs, vector control, and environmental modification. Despite intensive research, no malaria vaccine is commercially yet available. The vaccine farthest along in field testing is based on a single malaria antigen (circumsporozoite protein, CSP) and is not as effective as experimental radiation-attenuated whole-parasite vaccines. When immune responses to the protective irradiated parasite vaccines are analyzed, no single target antigen has been identified that explains the full extent of host immunity. Protection is thought to be strongly associated with interferon-gamma (IFN-$\gamma$) secretion by CD8+ T cell immunity during the liver stage of infection. This suggests that protective vaccines should be designed that are specifically capable of stimulating malaria antigen-specific CD8+ T cell responses. We developed a multi-antigen malaria vaccine, which is specifically designed to generate high levels of antigen-specific CD8 CTL that localize to the liver. Secreted gp96-Ig chaperoning *Plasmodium falciparum* (Pf) sporozoite proteins are efficiently taken up and cross-presented by activated DC via MHC I to CD8 CTL, thereby stimulating an avid, antigen-specific, cytotoxic T cell response. The generation of a powerful, cytotoxic anti-sporozoite CD8 CTL response by the vaccine is expected to provide prophylactic immunity against malaria by killing infected liver cells, thereby preventing blood stage infection.

## 5.5 Summary

The concept of stimulating the body's immune response is the basis underlying vaccination. Secreted gp96-Ig vaccines act by initiating the innate immune response and activating antigen-presenting cells (APCs), thereby inducing a protective adaptive immune response to a tumor and pathogen antigens. Several key properties distinguish gp96-based vaccination strategies from all other peptide-, protein-, or DNA-based vaccines:

1. gp96 is efficient in antigen cross-presentation at physiologic concentrations of antigen (picogram range), so even endogenous amounts of gp96-antigenic peptide complexes secreted by vaccine cells induce robust CTL responses.
2. gp96 is itself a natural adjuvant since dendritic cells are activated by gp96 binding through CD91 and TLR receptors.

3. gp96 does not need to be given in a complex with an artificial adjuvant to stimulate an immune response.
4. Since peptides are presented on vaccine recipient's dendritic cells, it is possible to vaccinate with allogeneic cells containing the appropriate antigens and secreting gp96-Ig across MHC barriers.

Secreted gp96-Ig vaccines provide an exciting and avant-garde strategy for the development of much needed vaccines; data from clinical trials are now needed to confirm that gp96-Ig vaccines provide an effective new approach in man.

## References


1. Robbins JB, Schneerson R, Trollfors B, Sato H, Sato Y, Rappuoli R, Keith JM (2005) The diphtheria and pertussis components of diphtheria-tetanus toxoids-pertussis vaccine should be genetically inactivated mutant toxins. J Infect Dis 191:81–88
2. Robinson MS, Watts C, Zerial M (1996) Membrane dynamics in endocytosis. Cell 84:13–21
3. Johnston MI, Fauci AS (2007) An HIV vaccine—evolving concepts. N Engl J Med 356:2073–2081
4. Mcmichael AJ (2006) HIV vaccines. Annu Rev Immunol 24:227–255
5. Rappuoli R (2004) From Pasteur to genomics: progress and challenges in infectious diseases. Nat Med 10:1177–1185
6. Rappuoli R (2007) Bridging the knowledge gaps in vaccine design. Nat Biotechnol 25:1361–1366
7. Strbo N, Garcia-Soto A, Schreiber TH, Podack ER (2013a) Secreted heat shock protein gp96-Ig: next-generation vaccines for cancer and infectious diseases. Immunol Res 57:311–325
8. Moser M, Leo O (2010) Key concepts in immunology. Vaccine 28(Suppl 3):C2–13
9. Basu S, Binder RJ, Suto R, Anderson KM, Srivastava PK (2000) Necrotic but not apoptotic cell death releases heat shock proteins, which deliver a partial maturation signal to dendritic cells and activate the NF-kappa B pathway. Int Immunol 12:1539–1546
10. Chen W, Syldath U, Bellmann K, Burkart V, Kolb H (1999) Human 60-kDa heat-shock protein: a danger signal to the innate immune system. J Immunol 162:3212–3219
11. Lehner T, Bergmeier LA, Wang Y, Tao L, Sing M, Spallek R, van der Zee R (2000) Heat shock proteins generate beta-chemokines which function as innate adjuvants enhancing adaptive immunity. Eur J Immunol 30:594–603
12. Panjwani NN, Popova L, Srivastava PK (2002) Heat shock proteins gp96 and hsp70 activate the release of nitric oxide by APCs. J Immunol 168:2997–3003
13. Singh-Jasuja H, Scherer HU, Hilf N, Arnold-Schild D, Rammensee HG, Toes RE, Schild H (2000) The heat shock protein gp96 induces maturation of dendritic cells and down-regulation of its receptor. Eur J Immunol 30:2211–2215
14. Binder RJ, Anderson KM, Basu S, Srivastava PK (2000) Cutting edge: heat shock protein gp96 induces maturation and migration of CD11c+ cells in vivo. J Immunol 165:6029–6035
15. Zhou YJ, Messmer MN, Binder RJ (2014) Establishment of tumor-associated immunity requires interaction of heat shock proteins with CD91. Cancer Immunol Res 2:217–228
16. Bach FH, Geller RL, Nelson PJ, Panzer S, Gromo G, Benfield MR, Inverardi L, Podack ER, Witson JC, Houchins JP et al (1989) A "minimal signal-stepwise activation" analysis of functional maturation of T lymphocytes. Immunol Rev 111:35–57
17. Baker M, Podack ER, Levy RB (1995) Fas and perforin cytotoxic pathways are not the major effector mechanisms in allogeneic resistance to bone marrow. Ann N Y Acad Sci 770:368–369

18. Baker MB, Altman NH, Podack ER, Levy RB (1996) The role of cell-mediated cytotoxicity in acute GVHD after MHC-matched allogeneic bone marrow transplantation in mice. J Exp Med 183:2645–2656
19. Biesecker G, Podack ER, Halverson CA, Muller-Eberhard HJ (1979) C5b-9 dimer: isolation from complement lysed cells and ultrastructural identification with complement-dependent membrane lesions. J Exp Med 149:448–458
20. Blazar BR, Levy RB, Mak TW, Panoskaltsis-Mortari A, Muta H, Jones M, Roskos M, Serody JS, Yagita H, Podack ER, Taylor PA (2004) CD30/CD30 ligand (CD153) interaction regulates CD4+ T cell-mediated graft-versus-host disease. J Immunol 173:2933–2941
21. Fang L, Adkins B, Deyev V, Podack ER (2008) Essential role of TNF receptor superfamily 25 (TNFRSF25) in the development of allergic lung inflammation. J Exp Med 205:1037–1048
22. Hatfield SM, Kjaergaard J, Lukashev D, Schreiber TH, Belikoff B, Abbott R, Sethumadhavan S, Philbrook P, Ko K, Cannici R, Thayer M, Rodig S, Kutok JL, Jackson EK, Karger B, Podack ER, Ohta A, Sitkovsky MV (2015) Immunological mechanisms of the antitumor effects of supplemental oxygenation. Sci Transl Med 7:277ra30
23. Kagi D, Ledermann B, Burki K, Seiler P, Odermatt B, Olsen KJ, Podack ER, Zinkernagel RM, Hengartner H (1994) Cytotoxicity mediated by T cells and natural killer cells is greatly impaired in perforin-deficient mice. Nature 369:31–37
24. Lichtenheld MG, Olsen KJ, Lu P, Lowrey DM, Hameed A, Hengartner H, Podack ER (1988) Structure and function of human perforin. Nature 335:448–451
25. Oizumi S, Strbo N, Pahwa S, Deyev V, Podack ER (2007) Molecular and cellular requirements for enhanced antigen cross-presentation to CD8 cytotoxic T lymphocytes. J Immunol 179:2310–2317
26. Podack ER (1985) The molecular mechanism of lymphocyte-mediated tumor cell lysis. Immunol Today 6:21–27
27. Podack ER, Konigsberg PJ (1984) Cytolytic T cell granules. Isolation, structural, biochemical, and functional characterization. J Exp Med 160:695–710
28. Podack ER, Young JD, Cohn ZA (1985) Isolation and biochemical and functional characterization of perforin 1 from cytolytic T-cell granules. Proc Natl Acad Sci U S A 82:8629–8633
29. Raez LE, Cassileth PA, Schlesselman JJ, Sridhar K, Padmanabhan S, Fisher EZ, Baldie PA, Podack ER (2004) Allogeneic vaccination with a B7.1 HLA-A gene-modified adenocarcinoma cell line in patients with advanced non-small-cell lung cancer. J Clin Oncol 22:2800–2807
30. Raez LE, Fein S, Podack ER (2005a) Lung cancer immunotherapy. Clin Med Res 3:221–228
31. Raez LE, Rosenblatt JD, Podack ER (2006) Present and future of lung cancer vaccines. Expert Opin Emerg Drugs 11:445–459
32. Schreiber TH, Podack ER (2013) Immunobiology of TNFSF15 and TNFRSF25. Immunol Res 57:3–11
33. Strbo N, Oizumi S, Sotosek-Tokmadzic V, Podack ER (2003) Perforin is required for innate and adaptive immunity induced by heat shock protein gp96. Immunity 18:381–390
34. Strbo N, Pahwa S, Kolber MA, Gonzalez L, Fisher E, Podack ER (2010) Cell-secreted Gp96-Ig-peptide complexes induce lamina propria and intraepithelial CD8+ cytotoxic T lymphocytes in the intestinal mucosa. Mucosal Immunol 3:182–192
35. Strbo N, Vaccari M, Pahwa S, Kolber MA, Doster MN, Fisher E, Gonzalez L, Stablein D, Franchini G, Podack ER (2013b) Cutting edge: novel vaccination modality provides significant protection against mucosal infection by highly pathogenic simian immunodeficiency virus. J Immunol 190:2495–2499
36. Tschopp J, Muller-Eberhard HJ, Podack ER (1982) Formation of transmembrane tubules by spontaneous polymerization of the hydrophilic complement protein C9. Nature 298:534–538
37. Yamazaki K, Nguyen T, Podack ER (1999) Cutting edge: tumor secreted heat shock-fusion protein elicits CD8 cells for rejection. J Immunol 163:5178–5182
38. Raez LE, Santos ES, Mudad R, Podack ER (2005b) Clinical trials targeting lung cancer with active immunotherapy: the scope of vaccines. Expert Rev Anticancer Ther 5:635–644

39. Schreiber TH, Deyev VV, Rosenblatt JD, Podack ER (2009) Tumor-induced suppression of CTL expansion and subjugation by gp96-Ig vaccination. Cancer Res 69:2026–2033
40. Strbo N, Vaccari M, Pahwa S, Kolber MA, Fisher E, Gonzalez L, Doster MN, Hryniewicz A, Felber BK, Pavlakis GN, Franchini G, Podack ER (2011) Gp96 SIV Ig immunization induces potent polyepitope specific, multifunctional memory responses in rectal and vaginal mucosa. Vaccine 29:2619–2625
41. Oizumi S, Deyev V, Yamazaki K, Schreiber T, Strbo N, Rosenblatt J, Podack ER (2008) Surmounting tumor-induced immune suppression by frequent vaccination or immunization in the absence of B cells. J Immunother 31:394–401
42. Strbo N, Podack ER (2008) Secreted heat shock protein gp96-Ig: an innovative vaccine approach. Am J Reprod Immunol 59:407–416
43. Dollins DE, Immormino RM, Gewirth DT (2005) Structure of unliganded GRP94, the endoplasmic reticulum Hsp90. Basis for nucleotide-induced conformational change. J Biol Chem 280:30438–30447
44. Immormino RM, Dollins DE, Shaffer PL, Soldano KL, Walker MA, Gewirth DT (2004) Ligand-induced conformational shift in the N-terminal domain of GRP94, an Hsp90 chaperone. J Biol Chem 279:46162–46171
45. Peng P, Menoret A, Srivastava PK (1997) Purification of immunogenic heat shock protein 70-peptide complexes by ADP-affinity chromatography. J Immunol Methods 204:13–21
46. Udono H, Srivastava PK (1993) Heat shock protein 70-associated peptides elicit specific cancer immunity. J Exp Med 178:1391–1396
47. Linderoth NA, Popowicz A, Sastry S (2000) Identification of the peptide-binding site in the heat shock chaperone/tumor rejection antigen gp96 (Grp94). J Biol Chem 275:5472–5477
48. Sastry S, Linderoth N (1999) Molecular mechanisms of peptide loading by the tumor rejection antigen/heat shock chaperone gp96 (GRP94). J Biol Chem 274:12023–12035
49. Wearsch PA, Nicchitta CV (1997) Interaction of endoplasmic reticulum chaperone GRP94 with peptide substrates is adenine nucleotide-independent. J Biol Chem 272:5152–5156
50. Linderoth NA, Simon MN, Hainfeld JF, Sastry S (2001a) Binding of antigenic peptide to the endoplasmic reticulum-resident protein gp96/GRP94 heat shock chaperone occurs in higher order complexes. Essential role of some aromatic amino acid residues in the peptide-binding site. J Biol Chem 276:11049–11054
51. Linderoth NA, Simon MN, Rodionova NA, Cadene M, Laws WR, Chait BT, Sastry S (2001b) Biophysical analysis of the endoplasmic reticulum-resident chaperone/heat shock protein gp96/GRP94 and its complex with peptide antigen. Biochemistry 40:1483–1495
52. Biswas C, Sriram U, Ciric B, Ostrovsky O, Gallucci S, Argon Y (2006) The N-terminal fragment of GRP94 is sufficient for peptide presentation via professional antigen-presenting cells. Int Immunol 18:1147–1157
53. Gidalevitz T, Biswas C, Ding H, Schneidman-Duhovny D, Wolfson HJ, Stevens F, Radford S, Argon Y (2004) Identification of the N-terminal peptide binding site of glucose-regulated protein 94. J Biol Chem 279:16543–16552
54. Ying M, Flatmark T (2006) Binding of the viral immunogenic octapeptide VSV8 to native glucose-regulated protein Grp94 (gp96) and its inhibition by the physiological ligands ATP and Ca2+. FEBS J 273:513–522
55. Melnick J, Dul JL, Argon Y (1994) Sequential interaction of the chaperones BiP and GRP94 with immunoglobulin chains in the endoplasmic reticulum. Nature 370:373–375
56. Randow F, Seed B (2001) Endoplasmic reticulum chaperone gp96 is required for innate immunity but not cell viability. Nat Cell Biol 3:891–896
57. Yang Y, Liu B, Dai J, Srivastava PK, Zammit DJ, Lefrancois L, Li Z (2007) Heat shock protein gp96 is a master chaperone for toll-like receptors and is important in the innate function of macrophages. Immunity 26:215–226
58. Suto R, Srivastava PK (1995) A mechanism for the specific immunogenicity of heat shock protein-chaperoned peptides. Science 269:1585–1588

59. Udono H, Srivastava PK (1994) Comparison of tumor-specific immunogenicities of stress-induced proteins gp96, hsp90, and hsp70. J Immunol 152:5398–5403
60. Riley JP, Rosenberg SA, Parkhurst MR (2001) Identification of a new shared HLA-A2.1 restricted epitope from the melanoma antigen tyrosinase. J Immunother 24:212–220
61. Schreiber TH, Raez L, Rosenblatt JD, Podack ER (2010) Tumor immunogenicity and responsiveness to cancer vaccine therapy: the state of the art. Semin Immunol 22:105–112
62. de Smet C, Lurquin C, de Plaen E, Brasseur F, Zarour H, de Backer O, Coulie PG, Boon T (1997) Genes coding for melanoma antigens recognised by cytolytic T lymphocytes. Eye (Lond) 11(Pt 2):243–248
63. Singhal S, Miller D, Ramalingam S, Sun SY (2008) Gene expression profiling of non-small cell lung cancer. Lung Cancer 60:313–324
64. Liyanage UK, Moore TT, Joo HG, Tanaka Y, Herrmann V, Doherty G, Drebin JA, Strasberg SM, Eberlein TJ, Goedegebuure PS, Linehan DC (2002) Prevalence of regulatory T cells is increased in peripheral blood and tumor microenvironment of patients with pancreas or breast adenocarcinoma. J Immunol 169:2756–2761
65. Liyanage UK, Goedegebuure PS, Moore TT, Viehl CT, Moo-Young TA, Larson JW, Frey DM, Ehlers JP, Eberlein TJ, Linehan DC (2006) Increased prevalence of regulatory T cells (Treg) is induced by pancreas adenocarcinoma. J Immunother 29:416–424
66. Beyer M, Schultze JL (2006) Regulatory T cells in cancer. Blood 108:804–811
67. Itoh K, Tilden AB, Balch CM (1986) Interleukin 2 activation of cytotoxic T-lymphocytes infiltrating into human metastatic melanomas. Cancer Res 46:3011–3017
68. Mellor AL, Munn DH (2008) Creating immune privilege: active local suppression that benefits friends, but protects foes. Nat Rev Immunol 8:74–80
69. Rabinovich GA, Gabrilovich D, Sotomayor EM (2007) Immunosuppressive strategies that are mediated by tumor cells. Annu Rev Immunol 25:267–296
70. Azoury SC, Straughan DM, Shukla V (2015) Immune checkpoint inhibitors for cancer therapy: clinical efficacy and safety. Curr Cancer Drug Targets 15:452–462
71. Fife BT, Bluestone JA (2008) Control of peripheral T-cell tolerance and autoimmunity via the CTLA-4 and PD-1 pathways. Immunol Rev 224:166–182
72. Okazaki T, Chikuma S, Iwai Y, Fagarasan S, Honjo T (2013) A rheostat for immune responses: the unique properties of PD-1 and their advantages for clinical application. Nat Immunol 14:1212–1218
73. Bolli E, Quaglino E, Arigoni M, Lollini PL, Calogero R, Forni G, Cavallo F (2011) Oncoantigens for an immune prevention of cancer. Am J Cancer Res 1:255–264
74. Wolchok JD, Chan TA (2014) Cancer: antitumour immunity gets a boost. Nature 515:496–498
75. Sitkovsky M, Lukashev D (2005) Regulation of immune cells by local-tissue oxygen tension: HIF1 alpha and adenosine receptors. Nat Rev Immunol 5:712–721
76. Ohta A, Madasu M, Subramanian M, Kini R, Jones G, Chouker A, Ohta A, Sitkovsky M (2014) Hypoxia-induced and A2A adenosine receptor-independent T-cell suppression is short lived and easily reversible. Int Immunol 26:83–91
77. Hatfield SM, Kjaergaard J, Lukashev D, Belikoff B, Schreiber TH, Sethumadhavan S, Abbott R, Philbrook P, Thayer M, Shujia D, Rodig S, Kutok JL, Ren J, Ohta A, Podack ER, Karger B, Jackson EK, Sitkovsky M (2014) Systemic oxygenation weakens the hypoxia and hypoxia inducible factor 1alpha-dependent and extracellular adenosine-mediated tumor protection. J Mol Med (Berl) 92:1283–1292
78. Akondy RS, Monson ND, Miller JD, Edupuganti S, Teuwen D, Wu H, Quyyumi F, Garg S, Altman JD, Del Rio C, Keyserling HL, Ploss A, Rice CM, Orenstein WA, Mulligan MJ, Ahmed R (2009) The yellow fever virus vaccine induces a broad and polyfunctional human memory CD8+ T cell response. J Immunol 183:7919–7930
79. Precopio ML, Betts MR, Parrino J, Price DA, Gostick E, Ambrozak DR, Asher TE, Douek DC, Harari A, Pantaleo G, Bailer R, Graham BS, Roederer M, Koup RA (2007) Immunization with vaccinia virus induces polyfunctional and phenotypically distinctive CD8(+) T cell responses. J Exp Med 204:1405–1416

80. Genesca M, Skinner PJ, Bost KM, Lu D, Wang Y, Rourke TL, Haase AT, Mcchesney MB, Miller CJ (2008a) Protective attenuated lentivirus immunization induces SIV-specific T cells in the genital tract of rhesus monkeys. Mucosal Immunol 1:219–228
81. Genesca M, Skinner PJ, Hong JJ, Li J, Lu D, Mcchesney MB, Miller CJ (2008b) With minimal systemic T-cell expansion, CD8+ T cells mediate protection of rhesus macaques immunized with attenuated simian-human immunodeficiency virus SHIV89.6 from vaginal challenge with simian immunodeficiency virus. J Virol 82:11181–11196
82. Ramirez SR, Singh-Jasuja H, Warger T, Braedel-Ruoff S, Hilf N, Wiemann K, Rammensee HG, Schild H (2005) Glycoprotein 96-activated dendritic cells induce a CD8-biased T cell response. Cell Stress Chaperones 10:221–229
83. Schreiber TH, Wolf D, Bodero M, Gonzalez L, Podack ER (2012) T cell costimulation by TNFR superfamily (TNFRSF)4 and TNFRSF25 in the context of vaccination. J Immunol 189:3311–3318
84. Selinger C, Strbo N, Gonzalez L, Aicher L, Weiss JM, Law GL, Palermo RE, Vaccari M, Franchini G, Podack ER, Katze MG (2014) Multiple low-dose challenges in a rhesus macaque AIDS vaccine trial result in an evolving host response that affects protective outcome. Clin Vaccine Immunol 21:1650–1660
85. Rerks-Ngarm S, Pitisuttithum P, Nitayaphan S, Kaewkungwal J, Chiu J, Paris R, Premsri N, Namwat C, de Souza M, Adams E, Benenson M, Gurunathan S, Tartaglia J, Mcneil JG, Francis DP, Stablein D, Birx DL, Chunsuttiwat S, Khamboonruang C, Thongcharoen P, Robb ML, Michael NL, Kunasol P, Kim JH, Investigators M-T (2009) Vaccination with ALVAC and AIDSVAX to prevent HIV-1 infection in Thailand. N Engl J Med 361:2209–2220

# Chapter 6
# Regulation of the Extracellular Matrix by Heat Shock Proteins and Molecular Chaperones

**Natasha Marie-Eraine Boel and Adrienne Lesley Edkins**

**Abstract** The extracellular matrix (ECM) serves as a scaffold for cells within tissues and is composed of an intricate network of glycoproteins, growth factors and matricellular proteins which cooperatively function in cell processes such as migration, adhesion and wound healing. ECM morphology is constantly undergoing remodelling (synthesis, assembly and degradation) during normal cell processes and when deregulated may contribute to disease. Heat shock proteins (Hsps) are involved in regulating processes that determine the assembly and degradation of the ECM at multiple levels, in both normal and diseased states. These roles include mediating the activation of ECM-degrading enzymes, maintaining matrix stability and clearing aggregated/misfolded proteins. Hsp may serve as chaperones and receptors or have cytokine-like functions. In this chapter, we review how Hsp90, Hsp70, Hsp40 and a number of ER resident chaperones contribute to ECM regulation. The role of the non-Hsp chaperones, SPARC and clusterin in the ECM is also discussed.

**Keywords** Extracellular matrix · Chaperone · Hsp90 · Hsp70 · Hsp40 · sHsp

## 6.1 The Extracellular Matrix

A substantial portion of the volume of tissues is extracellular space occupied by an intricate network of macromolecules constituting the extracellular matrix (ECM). The ECM is the noncellular component of tissues and organs, which exists to provide essential physical and biochemical cues for the cell and is one of the most important regulators of cellular and tissue function in the body [1]. The ECM is composed of approximately 300 proteins [2] encompassing structural proteins,

N. M.-E. Boel · A. L. Edkins (✉)
Biomedical Biotechnology Research Unit, Department of Biochemistry and Microbiology, Rhodes University, Grahamstown, South Africa
e-mail: a.edkins@ru.ac.za



R. J. Binder, P. K. Srivastava (eds.), *Heat Shock Proteins in the Immune System*,
https://doi.org/10.1007/978-3-319-69042-1_6

**Table 6.1** Molecules constituting the extracellular matrix

| ECM protein | Function | Link with disease | References |
|---|---|---|---|
| Fibronectin | Structural glycoprotein. Roles in cell migration, growth, differentiation | Fibrosis, tumour progression | [3–5] |
| Collagen | Structural glycoprotein. Cell-ECM links, tissue rigidity and strength. Several types exist with distinct functions | Fibrosis, tumour progression, cardiovascular disease | [4, 6–9] |
| Laminin | Structural glycoprotein. Cell adhesion, migration, differentiation | Fibrosis, tumour progression | [4, 10] |
| Vitronectin | Glycoprotein associated with loose connective tissue and involved in wound healing and blood coagulation | Fibrosis, tumour progression | [155–157] |
| Elastin | Provides structural integrity mainly in connective tissues and arteries | Cardiovascular disease, cancer | [3, 158] |
| Dentin matrix protein 1 | Regulates nucleation of hydroxyapatite in bone ECM | Osteoporosis | [159, 160] |
| SPARC/ osteonectin | Non-structural. Mediates cell-matrix interactions and collagen biosynthesis | Arthritis, cancer, diabetes, osteoporosis | [11, 12, 161] |
| Thrombospondin | Non structural. Cell-matrix interactions | Angiogenesis | [158, 162] |
| Osteopontin | Non-structural. Ca-binding glycoprotein involved in attachment of cells in mineralized bone matrix | Osteoporosis | [159, 163] |
| Tenascin | Non-structural. Mediates inflammatory processes mainly in connective tissues | Tumour progression, inflammation | [158, 164] |

glycoproteins, growth factors and matricellular proteins (a partial list is presented in Table 6.1). These ECM molecules are divided into two subgroups: the basement membrane (BM) which underlies epithelial cells and interstitial/stromal ECM which constitutes the intercellular spaces. The BM is composed largely of laminins, type IV collagen and proteoglycans [6], whilst the stromal ECM includes collagens type I, II and III, fibronectin (FN), vitronectin (VN) and elastin [13]. Each tissue in animals has a specific type of ECM; the ECM of bone tissue is comprised of collagen fibres and bone mineral, whilst blood plasma constitutes the ECM of blood. Fibroblasts are the major cells responsible for synthesising and maintaining the ECM in connective tissue, whilst chondrocytes and osteoblasts are responsible for cartilage and bone ECM formation, respectively [14].

The ECM plays important roles in structural support and cell signalling and contains proteins and growth factors involved in regulating cell proliferation, migration, polarity and survival [1, 15], where it also plays important roles in the tumour microenvironment. The interaction of cells with the ECM is particularly important for regulating these processes. Cell-ECM adhesion, as well as the bidirectional communication between the ECM and the actin cytoskeleton, is mediated by ECM receptors [3]. These include transmembrane integrins, discoidin domain receptors

(DDRs), CD44, syndecans and urokinase-type plasminogen activator receptor (uPAR), all of which are important in inducing biomechanical signals, cell adhesion and migratory functions [6, 16]. The importance of the ECM is made evident by the large array of diseases that may arise from abnormalities in the ECM, including autoimmune and inflammatory diseases, cancer and atherosclerosis [17]. The matricellular proteins are nonstructural modulators of extracellular signals and are presumed to assist in providing a linkage between the ECM and cell surface receptors. These include thrombospondins, SPARC (secreted protein acidic and rich in cysteine), osteopontins and tenascin [18]. Many of these ECM proteins contain multiple domains with specific binding sites for macromolecules or receptors [1].

The ECM is highly dynamic and is constantly being remodelled to accommodate its variety of functions, a process which occurs in both physiological and pathological cases [19, 20]. Components of the ECM are degraded by matrix-degrading enzymes such as matrix metalloproteinases (MMPs), serine and threonine proteases, heparanases, cathepsins and a disintegrin and metalloproteinase with thrombospondin motifs (ADAMTs) [21]. This degradation of the ECM is tightly controlled, and changes in matrix homeostasis can have detrimental effects on essential cellular processes [22]. The ECM can be remodelled by proteolytic cleavage of ECM molecules by extracellular proteases, tensile stress or in response to signals transmitted by ECM receptors or growth factors [7, 20, 23]. Tissue homeostasis requires a balance between ECM synthesis and degradation. Perturbing homeostasis by loss of function mutations or modifications of ECM molecules and excessive deposition or removal of ECM components results in progression of various disease states including fibrosis, cancer and other developmental abnormalities [6, 7]. The ECM is of great importance in the tumour microenvironment as it may serve to promote or prevent tumour progression [8]. Cancer development and progression requires constant remodelling of the ECM which it does by regulating various pathways in cancer cells. For example, increased expression and deposition of collagen and fibronectin promote breast tumour rigidity resulting in poor prognosis [24, 25]. The recognised importance of the ECM in mediating disease has increased targeted therapies of the ECM [3]. The ECM is dynamic and constantly being remodelled in response to cellular cues. A substantial portion of the ECM is comprised of extracellular proteins, whose structure needs to be maintained during these dynamic remodelling processes. This is achieved in part by the activity of Hsp, which regulates the activity of some ECM proteins in both intracellular and extracellular compartments.

## 6.2 Heat Shock Proteins (Hsps) and Molecular Chaperones

Heat shock proteins (Hsps) are ubiquitous, highly conserved proteins which play similar roles in both eukaryotes and prokaryotes. These families of proteins are broadly categorised according to their molecular weight, small Hsps, Hsp40, Hsp60, Hsp70, Hsp90 and Hsp100, each having distinct functions within the cell [26].

Under conditions of stress, such as heat or hypoxia, Hsps accumulate in the cell to prevent protein misfolding and aggregation [27]. Hsps are catalysts for protein folding, and many act in ATP-dependent cycles together with a host of accessory proteins to regulate multiple processes, including folding, proteolysis, aggregation and translocation, which are required for cellular homeostasis. Certain Hsps function as molecular chaperones to facilitate the correct folding of nascent polypeptides and maintain protein homeostasis. Molecular chaperones are well-known ubiquitous proteins involved in maintaining proteostasis by functionally assisting nascent peptide folding or refolding of denatured or misfolded proteins [27]. Chaperones act at the level of protein folding to maintain correctly folded and active states of intracellular proteins, assembly of complexes and protein transport [28, 29]. The same is true for proteins requiring export from the cell, including cytokines, extracellular matrix (ECM) proteins and receptors. Due to the high degree of stress and interactions of extracellular proteins which might occur in the extracellular environment, Hsps are also responsible for stabilising misfolded and/or aggregated proteins extracellularly [30]. Numerous studies have focussed on identifying the roles of intracellular Hsp as molecular chaperones, but much less is known about the extracellular pool of Hsp and how they function in the extracellular space. Some have proposed that extracellular chaperones are derived from the intracellular counterparts via apoptosis [31], necrosis [32] or secretion by exosomes [33, 34]. Functional roles have been postulated for extracellular Hsp chaperones including cancer cell invasion and migration [35, 36] facilitating the immune response [37–41] and in the pathogenicity of bacterial and viral infections [42]. It is known that uncontrolled protein folding or aggregation can lead to protein deposition disorders (PDD), many of which are associated with extracellular protein deposits that often contain chaperones [43]. Also, given the highly stressful and oxidising environment of the extracellular space [44, 45], there would appear to be a clear requirement for extracellular Hsp to control the stability of extracellular proteins and prevent protein aggregation. However, there is still debate as to whether Hsp can function as bona fide chaperones in the extracellular milieu or whether they rather fulfil a cytokine-like role in this compartment. Below we describe the role of some Hsp and chaperones and how they may contribute to the regulation of the ECM, at either an intracellular or extracellular level.

## 6.3 Regulation of the ECM by Hsp90

Cytosolic Hsp90 is one of the most abundant molecular chaperones in eukaryotes, comprising 1–2% of the total protein content within cells and increasing to 4–6% under conditions of stress [46]. There are two cytosolic isoforms of Hsp90 (Hsp90α and Hsp90β) as well as organelle resident isoforms (Grp94 in the endoplasmic reticulum/ER and TRAP1 in the mitochondrion). Hsp90 is responsible for maintaining the active conformation of over 300 intracellular client proteins and is a central component of the network of molecular chaperones in the cell which cooperates with Hsp70 and other co-chaperones and cofactors to regulate the folding, stability and activity of client proteins [47–49].

### 6.3.1 Extracellular Hsp90 and the ECM

Hsp90 can exist as both intracellular and extracellular forms. In 1986, Ullrich and colleagues identified Hsp90 on the surface of mouse cells [50]. Since then Hsp90 has been detected by several groups in the extracellular space and on the surface of various cell types including fibrosarcoma, neuronal cells [35, 51, 52] and breast cancer [53]. The term "extracellular" is often used interchangeably to describe both membrane-bound and extracellular soluble forms of Hsp90. It is not known by what mechanism Hsp90 localises to the extracellular space, but since these proteins lack a secretory signal sequence, it must follow an alternative pathway to that of the canonical Golgi transport secretory pathway [33]. Various Hsps as well as co-chaperones such as Hop and p23 have been found on the plasma membrane and in the extracellular space [34, 54, 55].

Fibronectins are glycoproteins that allow for cells to move through the ECM by creating cell-ECM connections together with collagens and cell surface integrins. Fibronectin (FN) is secreted by cells in an unfolded, soluble form and upon binding of integrins is able to form insoluble dimers to create a meshwork of FN fibres [56, 57]. The deposition of ECM molecules including collagen and thrombospondin has been shown to be dependent upon the presence of FN fibrils [58]. FN plays a major role in the stability and organisation of the ECM, although the exact processes that regulate FN catabolism are not well understood [59, 60].

Hunter and colleagues reported FN as a novel interacting protein of extracellular Hsp90 and suggested a direct role for Hsp90 in FN matrix stability and remodelling [53] (Table 6.2). Surface-associated Hsp90 and extracellular soluble Hsp90β were identified in breast cancer cell lines including Hs578T, MDA-231 and MCF-7 cells, and a complex containing Hsp90 and FN was identified by immunoprecipitation and tandem mass spectrometry. The authors further showed the direct binding of FN and Hsp90 in vitro and colocalisation of these proteins in breast cancer cell lines [53]. Addition of exogenous Hsp90β in Hs578T cells increased the formation of extracellular FN matrix, whilst knockdown of Hsp90α or Hsp90β decreased the proportion of extracellular FN matrix, suggesting a role for extracellular Hsp90 in FN fibril formation. Upon inhibition of Hsp90 by novobiocin (NOV) but not geldanamycin, the FN matrix was observed to become unstable and degraded by a receptor-mediated endocytic mechanism (Fig. 6.1) [53]. The exact mechanism by which Hsp90 influenced the FN matrix is presently unknown. It is not clear whether extracellular Hsp90 acted as a chaperone, cytokine or receptor for FN during internalisation or whether this Hsp90-mediated turnover of FN requires an additional receptor. We have preliminary unpublished evidence to suggest the latter. We identified a putative extracellular complex that exists between FN, Hsp90 and low-density lipoprotein receptor-related protein 1 (LRP1). Internalisation of FN upon inhibition

**Table 6.2** Overview of some chaperones involved in regulation of the ECM

| Chaperone | Location | Function in ECM | References |
|---|---|---|---|
| Hsp90 | Intra- and extracellular | Regulates turnover of FN | [53] |
| | Intra- and extracellular | Regulation of proteases for degradation of ECM | [52] |
| Grp94 | ER, extracellular | Regulates processing of integrins<br>Mediates cell signalling at the cell surface to promote cell motility | [9, 61] |
| Hsp90, Hop, Hsp70, p23 | Intra- and extracellular | Forms a complex to assist in activation of proteases | [52] |
| Hsp70 | Intra- and extracellular | Activates cytokines contributing to accumulation of ECM proteins | [62] |
| Hsp47 | ER, extracellular | Procollagen maturation and collagen fibril processing in ECM | [63–65] |
| Grp78/Bip | ER, extracellular | Endocytoses DMP1 in bone matrix.<br>Binds cell surface receptors mediating ECM degradation<br>Regulates transport of ECM proteins in ER | [10, 66, 67] |
| Hsp25/Hsp27 | Cytoplasm, ER | Trafficking of mature, active aggrecan to the cell surface | [68] |
| MRJ/DNAJB6 | Cytoplasm | Regulates uPAR-dependent cell adhesion to VN | [69, 70] |
| Clusterin | Extracellular | Clearing aggregates associated with protein deposition disorders | [71, 72] |
| SPARC | Extracellular | Stabilizes procollagen<br>Binds various ECM proteins<br>Regulates levels of matrix metalloproteinases | [124, 142] |

of Hsp90 by NOV appears to require the presence of LRP1 (unpublished observations). Given that both FN and Hsp90 bind LRP1, it may be that NOV mediated its effects on FN either by modulating the Hsp90-FN-LRP1 complex to promote LRP1-mediated endocytic clearance of extracellular FN or by activation of a signalling pathway due to a change in the Hsp90-LRP1 interaction or indeed a combination of both mechanisms (unpublished observations).

Hsp90 has also been shown to regulate the function of FN-binding integrins in the ECM (Table 6.2). Inhibition of cell surface Hsp90, with the small molecule, cell-impermeable N-terminal inhibitor DMAG-N-oxide, suppressed β1 integrin-mediated cell invasion in Matrigel invasion assays in T24 human bladder cancer, PC3M prostate cancer and B16 murine melanoma cancer cell lines [73]. FN induces the association of integrins and c-Src at focal adhesion points, a key component in the cell migration process [74]. Tsutsumi and colleagues analysed focal adhesion assembly by immunoprecipitation of β1-integrin on FN coated and uncoated surfaces to determine that DMAG-N-oxide reduced c-Src binding to integrin. Disrupting the FN-stimulated interaction of β1-integrin with c-Src by the extracellular Hsp90 inhibitor reduced cell adhesion to the ECM and decreased cell motility

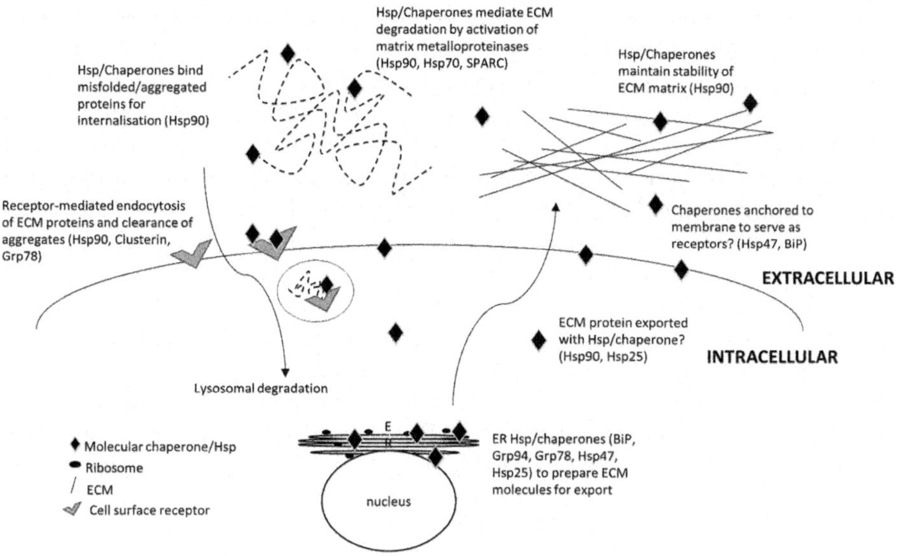


**Fig. 6.1** Schematic illustration of intracellular and extracellular regulation of the ECM by Hsp/molecular chaperones. Soluble matrix molecules are modified and processed in the ER by ER molecular chaperones and enzymes and subsequently transported to the cell surface presumably bound to chaperones for secretion. Extracellularly, ECM molecules assemble into polymeric complexes to form a structural scaffold or associate with the ECM assisted in part by extracellular Hsp/chaperones which may serve as receptors to mediate this cell-ECM association. Proteases activated by Hsp/chaperones degrade the scaffold in response to normal tissue turnover or wound healing, and the matrix molecules are internalised often via receptor-mediated endocytosis either by binding to cell surface receptors or chaperones or both. Misfolded and/or aggregated matrix molecules are bound by extracellular Hsp/chaperones for clearance from the ECM and subsequent intracellular degradation

[73]. Again, the mechanism by which Hsp90 reduced integrin-c-Src binding remains undefined.

In addition to FN, extracellular Hsp90 has also been shown to regulate ECM proteases (Table 6.2). In a proteomic screen for extracellular proteins that are important in cell invasion, extracellular Hsp90α was identified to interact with and activate MMP2 [52]. MMPs are central players in cell migration and invasion due to their ability to digest ECM components and cleave cell adhesive contacts [75]. Co-immunoprecipitation revealed a complex of Hsp90 and co-chaperones including Hsp70, Hsp40, Hop and p23, which assisted Hsp90α in activating MMP2 extracellularly in MDA-MB-231 breast cancer cells [52]. Using zymography, the authors showed that in the presence of this complex, MMP2 activation was enhanced by 33% in an ATP-independent manner and was able to promote cell migration [35, 52]. However, the mechanism of activation by this extracellular complex is still unclear, and it is unknown whether extracellular Hsp90α acts as a chaperone or cytokine in its role in activating MMP2.

In HT-1080 fibrosarcoma cells, plasmin activation assays demonstrated the ability of extracellular Hsp90α to activate a second extracellular protease, plasmin, by converting it from its precursor, plasminogen, to the active plasmin in much the same way as that of MMP2 [33]. In the presence of DMAG-N-oxide, there was a 32% decrease in activated plasmin. Using transwell migration assays, inhibition of extracellular Hsp90α decreased tumour cell migration compared to control treated cells. Cell migration is dependent in part on proteolysis of the ECM, and these data suggest that plasmin may be contributing to cell migration via remodelling of the ECM and highlight the important regulatory role for Hsp90 in this process [33]. McCready and colleagues identified a cohort of extracellular Hsp90α-interacting proteins by mass spectrometry, most of which were in their inactive, precursor forms and which they propose are activated by extracellular Hsp90α. They suggest that the potential role for extracellular Hsp90α may be through the appropriate activation of these proteins which then contribute to cell migration and invasion by enhancing remodelling of the ECM (Fig. 6.1) [33]. However, the mechanism by which Hsp90 activates these extracellular proteases remains undefined, although McCready and colleagues speculate that it may involve the proteolytic processing of inactive, precursor forms of these pro-invasive proteins [52, 76].

### *6.3.2 ER-Resident Hsp90, Grp94 and the ECM*

Proteins destined for secretion from cells are folded in the endoplasmic reticulum (ER) where a variety of chaperones and enzymes act to ensure protein quality control. The environment in the ER has been likened to that of the extracellular space as both of these are highly oxidising environments [44, 77]. During protein synthesis, various ER Hsps, including Grp94, assist in the folding and translocation of various membrane and secretory proteins, including components of the ECM [77, 78] (Table 6.2).

Grp94, an ER homolog of Hsp90, is specifically required for processing of several integrins in haematopoietic stem cells (HSC) which are essential for mediating the interaction of these cells with the stem cell niche and ECM [79]. Grp94 knockout HSCs in adult bone marrow were unable to express α4 integrin since Grp94 is essential for the processing and proper folding of α4 integrin [80]. Grp94 null HSCs consequently showed reduced binding to FN at the cell surface thereby impairing HSC interactions with its surrounding niche [81]. Interestingly, Grp94 has also been localised to the cell surface of various cancers, suggesting a possible role in mediating cell signalling to promote cell proliferation and motility [61, 79].

Cartilage oligomeric matrix protein (COMP) is a glycoprotein of the thrombospondin family of extracellular matrix proteins found in the cartilage, ligaments and tendons [82]. Although it is a small component in the ECM, mutations of COMP are the main cause of pseudoachondroplasia (PSACH), a skeletal dysplasia [83, 84]. Using immunoprecipitation and fluorescence resonance energy transfer (FRET) analyses, Hecht and colleagues demonstrated that the ER chaperones calreticulin, protein disulphide isomerase (PDI) and Grp94 are involved in transporting normal COMP to the ECM and selectively retaining mutant COMP molecules in the ER of PSACH chondrocytes [85]. Previous studies by this group also showed two other ECM molecules, aggrecan and type IX collagen, to be retained in the ER in these cells [86]. These studies highlight the importance of the quality control mechanisms by Grp94 in preventing mutated and misfolded ECM proteins from being exported to the cell surface.

## 6.4 Regulation of the ECM by Hsp70

Hsp70 forms one of the major Hsp families and is responsible for binding client proteins in their inactive conformation and, together with Hsp40, participates in de novo and stress-related protein folding. For certain groups of client proteins, the Hsp70-Hsp40 system also prepares the client protein for transfer to the Hsp90 complex [47, 87].

### *6.4.1 Extracellular Hsp70 and the ECM*

Vascular smooth muscle cells (SMCs) produce a majority of the ECM proteins, the abnormal accumulation of which leads to cardiovascular diseases such as atherosclerosis [88]. SMCs have been found to overexpress the stress-inducible form of Hsp70 [89], and increased levels of Hsp70 have been observed in cardiovascular disease. Extracellular Hsp70 interacts with Toll-like receptors (TLR) inducing vascular pro-inflammatory cytokine synthesis which impacts on the ECM (Table 6.2). Amongst the known cytokines, transforming growth factor-beta (TGFβ) is one of the main regulators of ECM synthesis, cell growth, differentiation, migration and proliferation of SMCs [90, 91]. Studies by Gonzolas-Ramos [62] demonstrate that extracellular Hsp70 binds TLR4 in human aorta SMCs which activates ERK and JNK (regulators of TGFβ1 transcriptional activation) causing an abnormal increase in ECM protein synthesis (including FN and collagen I) as observed by immunoblotting. They further showed that the increase in FN protein synthesis was inhibited upon TGFβ1 inhibition with a blocking antibody. Consistent with this observation, siRNA knockdown of TLR4, with which extracellular Hsp70 interacted, rendered SMCs unable to induce synthesis of TGFβ1. Since TGFβ is a major regulator of collagen and FN deposition [92], this highlights a role for extracellular Hsp70 as a pro-fibrotic regulator of ECM synthesis and induces structural changes in the vascular walls of SMCs characteristic of atherosclerosis [62, 88]. Hsp70 thus potentially plays dual roles in atherosclerosis as both a chaperone [89] and a cytokine [93].

### 6.4.2 *ER-Resident Hsp70, BiP and the ECM*

BiP, also referred to as Grp78, is part of the Hsp70 family of chaperones which primarily functions in the ER for folding and assembly of membrane and secretory proteins (Fig. 6.1 and Table 6.2). Apart from its function as a molecular chaperone in the ER, BiP has also been found as a secreted protein [94], where it may possess extracellular functions in regulating inflammatory cytokines [95], and at the cell surface where it serves as a signalling receptor for $\alpha_2$ macroglobulin ($\alpha_2$M) [96] and facilitates entry of viruses [40].

Ravindran and colleagues identified BiP at the cell surface of mouse preosteoblasts and mouse primary calvarial osteoblast cells [66]. Although not much is known about the function and expression of BiP in the ECM, the authors demonstrated BiP could bind and endocytose type I collagen [97] and dentin matrix protein 1 (DMP1), a protein in bone ECM responsible for regulating hydroxyapatite [66] (Table 6.2). Using a solid-phase binding assay to demonstrate the interaction of native BiP secreted from the cell with type I collagen, the authors presented evidence for a regulatory role of this ER chaperone in the formation of mineralized matrix. BiP was shown to initiate calcium phosphate nucleation in vitro, an important process in maintaining calcium ions in the ECM during mineralized matrix formation [66].

Co-immunoprecipitation and immunofluorescence revealed β1-integrins and BiP in a complex at the cell surface of HT-29 and SW480 colon cancer cell lines, as well as the colocalisation of uPAR with integrins and BiP [67]. This colocalisation of uPAR and BiP is thought to increase levels of the matrix-degrading protease, plasmin, by promoting the close association of uPA and plasminogen. This causes the conversion of plasminogen to plasmin which enhances degradation of the ECM. Increased levels of BiP at the cell surface have been observed in metastatic cancer cells, where it reportedly bridges with β1-integrin and uPAR to mediate degradation of the ECM and promote invasion in Matrigel transwell assays [67]. Considering integrins cluster at the leading edge of migrating cells [98], and uPAR associates with integrins to promote cancer progression, the interaction of BiP with integrins and uPAR should directly affect cell-matrix adhesion to facilitate ECM degradation by activating the uPAR protease system, thereby promoting cell invasion.

## 6.5 Regulation of the ECM by Hsp40

The human genome encodes for over 30 distinct Hsp40 members [99–101] also referred to as DNAJ proteins [102]. Hsp40 proteins have a characteristic J-domain at its N-terminus, and members of this family are divided into types I, II and III (or DNAJ A, B and C) [103]. The Hsp40 family is ubiquitously expressed, and its main function is to regulate the ATP-dependent binding of substrates to Hsp70 [47, 87], although certain Hsp40 isoforms have also been shown to have independent chaperone activity [104, 105].

### *6.5.1 MRJ/DNAJB6 and the ECM*

Urokinase plasminogen activator receptor (uPAR) is an important regulator of ECM proteolysis and cell-ECM interactions and signalling [106]. uPAR is involved in numerous processes including wound healing and tumour progression [106, 107]. The urokinase plasminogen activator (uPA) binds to uPAR converting plasminogen to plasmin which actively digests ECM molecules and further activates MMPs for ECM proteolysis [108, 109]. uPAR also interacts with integrins, VN, cytokeratin and epidermal growth factor receptor (EGFR) to activate cell signalling pathways [107, 110]. uPAR-dependent cell adhesion to VN in the ECM is important in wound healing, immune response and tissue remodelling, and this interaction has been described on the surface of endothelial and U93 cells [111]. Lin and colleagues demonstrate this uPAR/VN interaction to be, in part, regulated by the Hsp40 MRJ (also referred to as DNAJB6) [69]. MRJ (mammalian relative of DNAJ) is a class II member of the Hsp40 family and is an important co-chaperone of Hsp70 [32]. Lin and colleagues identified MRJ and Hsp70 to form a tripartite complex with uPAR by co-immunoprecipitation. This complex regulates uPAR-mediated cell migration and adhesion to VN in HEK293 human embryonic kidney and HCT116 human colon carcinoma cells [70]. Using wound healing and transwell assays in HCT116 cells, the authors showed that knockdown of Hsp70 and/or MRJ significantly decreased cell surface uPAR, resulting in reduced adhesion to VN thereby inhibiting cell migration. The authors propose a mechanism that suggests decreased cell surface levels of uPAR due to MRJ depletion cause a loss in expression of MMPs which abrogates ECM degradation, thereby preventing cell migration [70]. This highlights an important role of this tripartite complex in promoting matrix degradation. They further demonstrated using co-IPs that the interaction of uPAR is specific to Hsp70 and MRJ in HCT116 and MDA-MB-231 cell lines, respectively, and further speculate this interaction to occur in the cytoplasm. Little is known about the interaction of the MRJ/Hsp70 complex with uPAR and the exact mechanism of how it regulates uPAR-mediated adhesion to VN, but expression of these proteins have been shown to correlate with poor prognosis in breast cancer [112].

Studies by Mitra and colleagues also reported MRJ to be a major regulator of breast cancer metastasis by altering the transcription of key secreted ECM proteins involved in migration and invasion. Using qPCR and tandem mass spectrometry analyses, expression of MRJ in MDA-MB-435 cells was shown to decrease the expression of osteopontin and osteonectin amongst others [100]. Osteopontin is an important ECM glycoprotein involved in regulating adhesion and migration in various cancers. Osteonectin (also known as SPARC) is also part of the ECM and important in migration. Reports have demonstrated that osteopontin expression is significantly increased in melanomas [113] and is associated with poor clinical outcome in several cancer cells including breast, colon and ovarian cancers. Mitra and coworkers found an inversely proportional relationship between the transcript levels of osteopontin and MRJ in melanoma species using qPCR analysis [100, 114]. MRJ was demonstrated to form a complex with heat shock cognate protein, Hsc70, and protein phosphatase, PP2A, to dephosphorylate glycogen synthase kinase 3β (GSK3β). This dephosphorylation regulates Wnt/β-catenin signalling by activating GSK3β to downregulate osteopontin [114].

### 6.5.2 ER Resident Hsp40, Hsp47 and the ECM

Hsp47, sometimes called colligin, J6 or gp46, is a substrate-specific, ER-resident Hsp40 chaperone belonging to the serpin family of serine protease inhibitors [115], which binds exclusively to procollagens in the ER [116, 117]. Collagen is one of the most abundant proteins in mammalian tissues (about 25%), and its unique helical structure necessitates a specific chaperone system [118]. Hsp47 binds and chaperones many types of collagen including types I–IV, which extends to both fibrillar collagen and basement membrane collagen [119]. Hsp47 was first shown to interact with type IV collagen at the cell surface of mouse F9 embryonal carcinoma and teratocarcinoma-derived parietal endoderm cell lines, where it was presumed to be involved in collagen cross-linking [116]. Studies performed in Hsp47 knockout mice demonstrated a critical importance for this chaperone in the synthesis and maturation of procollagens as well as the extracellular collagen matrix [63, 120]. This was determined by immunohistological analysis of the Hsp47$^{-/-}$ fibroblasts which showed an observed decrease in collagen fibril accumulation and an increase in intracellular accumulation of procollagen, whilst levels of FN and laminin remained unaffected [120]. Procollagen is secreted from cells and is converted into collagen by the removal of the N- and C-terminal propeptides by procollagen metalloproteinases extracellularly, allowing for the assembly of collagen fibrils in the ECM [118]. Hsp47 transiently binds nascent procollagen peptides which enter the ER co-translationally and assists in the folding and stabilisation of procollagen (Fig. 6.1). Hsp47 then assists in the formation of triple helical procollagens mediated largely by enzymes within the ER, including PDI and peptidylprolyl isomerase (PPIase) as well as ER chaperones, Grp94, BiP, calnexin and calreticulin [121]. As the triple helical procollagen progresses from the ER to the *cis*-Golgi, Hsp47 dissociates from procollagen and is recycled back to the ER [63, 117]. Procollagen molecules associate to form aggregates which are secreted from the cell and in this form are proposed to facilitate extracellular fibril formation [122, 123]. Procollagens at the cell surface may be cleaved by specific proteases to allow for the formation of collagen fibrils in the ECM [123]. Extracellularly, various other protein interactions regulate collagen fibril formation such as those with FN, integrins and SPARC [124]. Interestingly, the expression of Hsp47 has been shown to correlate with that of collagen. In some cell types that do not produce any collagen, such as macrophages, lymphocytes, myeloid leukaemia and HEK293 embryonic kidney cells, Hsp47 is also lacking [117, 125].

Hsp47 may also exert a pathological role as has been described in studies showing a correlation between increased Hsp47 expression and development of solid tumours [64]. The *SERPINH1* gene encoding Hsp47 is amplified in various human cancers [126]. Increased expression and deposition of FN and collagen have been linked to enhanced tumour growth and invasion; however, not much is known about how these ECM proteins are regulated during cancer progression [24, 127]. Recent studies by Zhu and colleagues in breast cancer lines (Hs578T, MDA-MB-231, T4-2, BT549) showed the molecular chaperone Hsp47 to be a key protein in regulating the ECM gene network, whereby increased expression of Hsp47 was shown to promote cancer progression in part by enhancing secretion and deposition of collagen and FN [64]. The mRNA levels of Hsp47 and FN were significantly correlated in these breast cancer cell lines. Using gene co-expression network analysis, Zhu and colleagues identified Hsp47 (*SERPINH1*) as a central node in regulating the ECM transcription network and revealed that Hsp47 expression, regulated by microRNA (miR)-29, is activated during breast cancer development which increased expression of collagen I, collagen IV and FN levels. Several microRNAs, miR-29, miR-200 and miR-300, were found to be enriched in ECM network genes [64]. miR-29 has recently been identified to alter the tumour microenvironment through repression of the ECM transcription network [128] and may modulate Hsp47 expression by activating the TGFβ pathway [64]. Reduced miR-29 levels increased Hsp47 expression, thereby activating the ECM network in breast cancer. Xenograft assays showed that Hsp47 silencing significantly inhibited tumour growth in vivo due to reduced deposition of collagen and FN. Addition of exogenous FN to Hsp47 depleted breast cancer cells and was able to partially restore invasiveness. Considering Hsp47 knockdown reduced focal adhesions in breast cancer cells (a major regulator of FN deposition), it might be plausible that reduced tumour cell invasiveness in Hsp47$^{-/-}$ cells is also due to an indirect effect on FN deposition [64]. The expression of Hsp47 has also been identified on the surface of human oral squamous cell carcinomas and LL/2 murine epidermoid carcinoma cell lines where it can bind procollagen molecules [78]. Co-immunoprecipitation of cross-linked species in each of the human and murine carcinoma cell lines identified Hsp47 in association with collagens and the tetraspanin transmembrane protein CD9 [78]. CD9 allows Hsp47 to anchor to the cell membrane where it can bind the procollagen peptides and modulate tumour cell migration by acting as a serpin protein inhibitor or as a receptor for collagen [78, 129].

Most fibrotic diseases occur due to excessive accumulation of matrix proteins as a result of either uncontrolled synthesis and/or degradation [17]. Accumulation of collagen comprises the bulk of fibrotic mass, and, since Hsp47 has been shown to be upregulated in fibrotic diseases and is a collagen-specific chaperone, it provides a selective target for controlling fibrotic disease, provided one can specifically target Hsp47 [65]. A study by Sato and colleagues observed a reduction in collagen accumulation in liver cirrhosis following the delivery of Hsp47-specific siRNA packaged for targeted delivery to hepatic cells [130].

## 6.6 Regulation of the ECM by Small Heat Shock Proteins (sHsps)

Hsp25 is a murine homolog of Hsp27 belonging to the family of small heat shock proteins (sHsps), and it plays a role in regulating metastasis via the ECM [131]. The Hsp25 gene occurs throughout eukaryotes and the protein is constitutively expressed. Aggrecan is a secreted protein consisting of two globular domains, G1 and G3, and is a major component of cartilage ECM [132]. By transfection studies in Chinese hamster ovary (CHO) cells by Zheng and colleagues, Hsp25 was demonstrated to bind nascent G3 of aggrecan in the cytosol during a translocational pause, after which Hsp25 in complex with G3 entered the ER lumen where it assisted proper folding of this aggrecan domain. Interestingly, Hsp25 was not able to bind the G1 domain of aggrecan. Zheng and colleagues propose that following dissociation of Hsp25 from G3 in the ER, G3 acts as an "intramolecular chaperone" by binding the nascent G1 domain to assist in correct folding and formation of aggrecan core protein which is subsequently modified in the Golgi and secreted to the extracellular space where it forms part of the ECM [68, 133].

## 6.7 Non-Hsp Chaperones Regulating the ECM

### *6.7.1 Clusterin*

Clusterin was one of the first secreted proteins to be identified as a mammalian extracellular chaperone. It has similar chaperone activity to the small heat shock proteins (sHsps) in that its expression is induced by heat shock [134, 135], and it stabilises conformations of proteins [121, 136], although it cannot catalyse protein refolding [134]. Clusterin has been proposed to be responsible for regulating ECM protein deposition by clearing potentially pathological aggregates (including extracellular amyloid deposits associated with Alzheimer's disease) from the extracellular space by binding one of the LDL receptors and targeting degradation of aggregated proteins to lysosomes [71, 72]. Extracellular protein deposits are associated with numerous diseases including Creutzfeldt-Jakob disease, age-related macular degeneration, atherosclerosis, pseudoexfoliation syndrome and Alzheimer's disease [137]. The uptake of aggregated complexes and protein deposits may play an important role in PDD [138]. Clusterin therefore may play an important role in PDD by interacting with and chaperoning Aβ peptides in the extracellular space associated with Alzheimer's disease and assisting in Aβ peptide clearance via receptor-mediated endocytosis commonly by megalin (LRP2) or LRP1 [139–141].

### *6.7.2 SPARC/Osteonectin*

SPARC, also called osteonectin or BM-40, is a glycoprotein belonging to the matricellular group of proteins and has no structural role in the ECM but rather serves to mediate cell-matrix interactions [11, 142]. SPARC is abundantly expressed in tissues undergoing repair and remodelling (such as during foetal bone development and wound healing) and in pathologies such as cancer, arthritis and diabetes [11]. Osteonectin is one of the most abundant non-collagenous matrix proteins in bone [143]. It is synthesised by fibroblasts and macrophages at wound sites and may regulate the deposition and assembly of ECM proteins [144]. In bovine aortic endothelial cells, addition of exogenous SPARC was able to significantly decrease the synthesis of FN and thrombospondin [144]. In synovial fibroblasts, SPARC increased levels of various MMPs involved in tissue remodelling, highlighting a role for SPARC in regulating components of the ECM including proteases and protease inhibitors (e.g. PAI-1) that affect cell-ECM interactions [12, 145]. SPARC interacts with a variety of factors that modulate the ECM, one of which is TGFβ1 which has been shown to consequently regulate expression of SPARC in fibroblasts [142]. In a review by Martinek and colleagues, they propose SPARC to have chaperone activities similar to that of Hsp47 in binding and stabilising procollagen molecules and triple helices being transported from the ER [124]. Interestingly, SPARC binds to a domain on procollagen which serves as a binding site for several ligands such as integrins and FN [146] thereby influencing the activity of ECM molecules responsible for fibrillogenesis and remodelling.

## 6.8 Concluding Remarks

Regulation of the ECM by Hsps occurs at various levels within the cell both intracellularly and extracellularly (Fig. 6.1). The importance of the array of ER chaperones in regulating the ECM by ensuring only fully functional ECM molecules are secreted from the cell has been highlighted. Once outside, some of the Hsps and chaperones have been demonstrated to remain associated with ECM components which may highlight additional extracellular regulatory roles of these chaperones (Fig. 6.1). Some Hsp may regulate the ECM gene network by altering transcriptional activation of certain ECM molecules and/or cytokines which enhances or reduces ECM synthesis and deposition thereby modulating matrix dynamics. Further to this, Hsps have been demonstrated to regulate activation of various ECM-degrading enzymes and the clearance of unstable ECM molecules or aggregates. As presented here, it is not uncommon for molecular chaperones to exhibit concomitant intracellular and extracellular regulation of ECM proteins.

Unfortunately, particularly in the context of extracellular Hsp, we still do not have a clear understanding of the mechanisms by which many Hsps function in the ECM. For example, in the case of extracellular Hsp90 and Hsp70, it is not clear whether these chaperones are functioning as an extracellular chaperone or as a cytokine. It is well established that intracellular Hsp90 and Hsp70 chaperone function is dependent on the formation of multi-chaperone complexes driven by the availability of ATP [147, 148]. ATP is secreted into the extracellular environment by some cell types [149–151]. However, the concentration of ATP in the extracellular environment is considered to be below the threshold required for intracellular chaperones, suggesting that the extracellular functions may be independent of ATP. Indeed, extracellular chaperones like clusterin are ATP-independent [152]. However, the absence of a requirement for ATP in the case of Hsp90 is not consistent with the effect of DMAG-N-oxide on some extracellular Hsp90 functions [153], as this is an N-terminal, ATP-competitive inhibitor [73]. It is possible that Hsp70 and Hsp90 are not acting as traditional chaperones in this ATP-limited environment but rather fulfil a cytokine-like role by activating intracellular signalling pathways that culminate in the associated biological changes [154]. Indeed, extracellular Hsp90 has been shown to activate Akt signalling downstream of the receptor LRP1, and Hsp70 can activate JNK and ERK downstream of the Hsp70-TLR4 complex [4, 5].

It is clear that many questions remain unanswered. The challenge ahead lies in establishing mechanisms by which these chaperones are able to regulate the ECM in different environments. Given that evidence implicating the ECM in disease progression is increasing, studies aimed at targeting Hsp that regulates ECM homeostasis in pathology should be prioritised [17].

**Acknowledgements** This work is based on the research supported by the South African Research Chairs Initiative of the Department of Science and Technology, the National Research Foundation of South Africa (Grant No 98566), the South African Medical Research Council (SAMRC), the Cancer Association of South Africa (CANSA), and Rhodes University. NMEB was supported with a postgraduate fellowship from the National Research Foundation. The views expressed are those of the authors and should not be attributed to the DST, NRF, CANSA, SAMRC or Rhodes University. We apologise if we have inadvertently omitted any important contributions to the field.

## References

1. Hynes RO (2009) The extracellular matrix: not just pretty fibrils. Science 326(5957):1216–1219. Available from: http://www.ncbi.nlm.nih.gov/pubmed/19965464. Cited 28 Feb 2013
2. Hynes RO, Naba A (2012) Overview of the matrisome—an inventory of extracellular matrix constituents and functions. Cold Spring Harb Perspect Biol 4(1):a004903. Available from: http://www.pubmedcentral.nih.gov/articlerender.fcgi?artid=3249625&tool=pmcentrez&rendertype=abstract
3. Theocharis AD, Skandalis SS, Gialeli C, Karamanos NK (2015) Extracellular matrix structure. Adv Drug Deliv Rev 97:4–27. https://doi.org/10.1016/j.addr.2015.11.001
4. Tsen F, Bhatia A, O'Brien K, Cheng C-F, Chen M, Hay N et al (2013) Extracellular heat shock protein 90 signals through subdomain II and the NPVY motif of LRP-1 receptor to Akt1 and Akt2: a circuit essential for promoting skin cell migration in vitro and wound healing in vivo. Mol Cell Biol 33(24):4947–4959. Available from: http://www.pubmedcentral.nih.gov/articlerender.fcgi?artid=3889557&tool=pmcentrez&rendertype=abstract. Cited 17 July 2014

5. Hance MW, Nolan KD, Isaacs JS (2014) The double-edged sword: conserved functions of extracellular Hsp90 in wound healing and cancer. Cancers (Basel) 6:1065–1097
6. Frantz C, Stewart KM, Weaver VM (2010) The extracellular matrix at a glance the extracellular matrix at a glance. J Cell Sci 2010:4195–4200
7. Lu P, Takai K, Weaver VM, Werb Z (2011) Extracellular matrix degradation and remodeling in development and disease. Cold Spring Harb Perspect Biol 3:1–21
8. Egeblad M, Rasch MG, Weaver VM (2010) Dynamic interplay between the collagen scaffold and tumor evolution. Curr Opin Cell Biol 22(5):697–706. Available from: http://www.pubmedcentral.nih.gov/articlerender.fcgi?artid=2948601&tool=pmcentrez&rendertype=abstract. Cited 29 Oct 2015
9. Liu B, Dai J, Zheng H, Stoilova D, Sun S, Li Z (2003) Cell surface expression of an endoplasmic reticulum resident heat shock protein gp96 triggers MyD88-dependent systemic autoimmune diseases. Proc Natl Acad Sci U S A 100(26):15824–15829. Available from: http://www.pubmedcentral.nih.gov/articlerender.fcgi?artid=307652&tool=pmcentrez&rendertype=abstract
10. Lee AS (2014) Glucose-regulated proteins in cancer: molecular mechanisms and therapeutic potential. Nat Rev Cancer 14(4):263–276
11. Brekken RA, Sage EH (2000) SPARC, a matricellular protein: at the crossroads of cell-matrix communication. Matrix Biol 19:816–827
12. Bradshaw AD (2012) Diverse biological functions of the SPARC family of proteins. Int J Cell Biol 44(3):480–488
13. Huang G, Greenspan DS (2012) ECM roles in the function of metabolic tissues. Trends Endocrinol Metab 23(1):16–22. Available from: http://www.pubmedcentral.nih.gov/articlerender.fcgi?artid=3251694&tool=pmcentrez&rendertype=abstract. Cited 17 July 2014
14. Erler JT, Weaver VM (2009) Three-dimensional context regulation of metastasis. Clin Exp Metastasis 26(1):35–49
15. Lin QC, Bissell MJ (1993) Multi-faceted regulation of of cell differentiation by extracellular matrix. FASEB J 7:737–743
16. Bridgewater RE, Norman JC, Caswell PT (2012) Integrin trafficking at a glance. J Cell Sci 125(Pt 16):3695–3701. Available from: http://www.pubmedcentral.nih.gov/articlerender.fcgi?artid=3462077&tool=pmcentrez&rendertype=abstract. Cited 16 July 2014
17. Cox TR, Erler JT (2011) Remodeling and homeostasis of the extracellular matrix: implications for fibrotic diseases and cancer. Dis Model Mech 4(2):165–178. Available from: http://dmm.biologists.org/content/4/2/165
18. Bornstein P, Sage EH (2002) Matricellular proteins: extracellular modulators of cell function. Curr Opin Cell Biol 14(5):608–616. Available from: http://www.ncbi.nlm.nih.gov/pubmed/12231357
19. Bonnans C, Chou J, Werb Z (2014) Remodelling the extracellular matrix in development and disease. Nat Rev Mol Cell Biol 15(12):786–801. Available from: http://www.nature.com/doifinder/10.1038/nrm3904. Cited 21 Nov 2014
20. Daley WP, Peters SB, Larsen M (2008) Extracellular matrix dynamics in development and regenerative medicine. J Cell Sci 121(Pt 3):255–264. Available from: http://www.ncbi.nlm.nih.gov/pubmed/18216330. Cited 4 Mar 2013
21. Mott JD, Werb Z (2004) Regulation of matrix biology by matrix metalloproteinases. Curr Opin Cell Biol 16(5):558–564
22. Egeblad M, Werb Z (2002) New functions for the matrix metalloproteinases in cancer progression. Nat Rev Cancer 2(3):161–174. Available from: http://www.nature.com/doifinder/10.1038/nrc745
23. Davies KJ (2014) The complex interaction of matrix metalloproteinases in the migration of cancer cells through breast tissue stroma. Int J Breast Cancer 2014:839094. Available from: http://www.pubmedcentral.nih.gov/articlerender.fcgi?artid=3985306&tool=pmcentrez&rendertype=abstract

24. Provenzano PP, Inman DR, Eliceiri KW, Knittel JG, Yan L, Rueden CT et al (2008) Collagen density promotes mammary tumor initiation and progression. BMC Med 6:11. Available from: http://www.pubmedcentral.nih.gov/articlerender.fcgi?artid=2386807&tool=pmcentrez&rendertype=abstract
25. Levental KR, Yu H, Kass L, Lakins JN, Erler JT, SFT F et al (2010) Matrix crosslinking forces tumor progression by enhancing integrin signaling. Cell 139(5):891–906
26. Hartl FU, Hayer-Hartl M (2002) Molecular chaperones in the cytosol: from nascent chain to folded protein. Science 295:1852–1858. Available from: http://www.sciencemag.org/cgi/doi/10.1126/science.1068408
27. Hendrick J, Hartl U (1993) Molecular chaperone functions of heat-shock proteins. Annu Rev Biochem 62(1):349–384
28. Ellgaard L, Molinari M, Helenius A (1999) Setting the standards: quality control in the secretory pathway. Science 286(5446):1882–1888
29. Kim YE, Hipp MS, Bracher A, Hayer-Hartl M, Hartl FU (2013) Molecular chaperone functions in protein folding and proteostasis. Annu Rev Biochem 82:323–355. Available from: http://www.ncbi.nlm.nih.gov/pubmed/23746257. Cited 10 July 2014
30. Bukau B, Weissman J, Horwich A (2006) Molecular chaperones and protein quality control. Cell 125(3):443–451. https://doi.org/10.1016/j.cell.2006.04.014
31. Feng H, Zeng Y, Whitesell L, Katsanıs E (2001) Stressed apoptotic tumor cells express heat shock proteins and elicit tumor-specific immunity. Blood 97(11):3505–3512
32. Saito K, Dai Y, Ohtsuka K (2005) Enhanced expression of heat shock proteins in gradually dying cells and their release from necrotically dead cells. Exp Cell Res 310(1):229–236
33. McCready J, Sims JD, Chan D, Jay DG (2010) Secretion of extracellular hsp90α via exosomes increases cancer cell motility : a role for plasminogen activation. BMC Cancer 10(294):1–10
34. Hegmans JPJJ, Bard MPL, Hemmes A, Luider TM, Kleijmeer MJ, Prins J-B et al (2004) Proteomic analysis of exosomes secreted by human mesothelioma cells. Am J Pathol 164(5):1807–1815. https://doi.org/10.1016/S0002-9440(10)63739-X
35. Eustace BK, Sakurai T, Stewart JK, Yimlamai D, Unger C, Zehetmeier C et al (2004) Functional proteomic screens reveal an essential extracellular role for hsp90 alpha in cancer cell invasiveness. Nat Cell Biol 6(6):507–514. Available from: http://www.ncbi.nlm.nih.gov/pubmed/15146192. Cited 9 July 2013
36. Sottile J, Chandler J (2005) Fibronectin matrix turnover occurs through a caveolin-1-dependent process. Mol Biol Cell 16(2):757–768. Available from: http://www.pubmedcentral.nih.gov/articlerender.fcgi?artid=545909&tool=pmcentrez&rendertype=abstract. Cited 4 Mar 2013
37. Binder RJ, Anderson KM, Basu S, Srivastava PK (2000) Cutting edge: heat shock protein gp96 induces maturation and migration of CD11c+ cells in vivo. J Immunol 165(11):6029–6035. Available from: http://www.ncbi.nlm.nih.gov/pubmed/11086034
38. Binder RJ, Srivastava PK (2005) Peptides chaperoned by heat-shock proteins are a necessary and sufficient source of antigen in the cross-priming of CD8+ T cells. Nat Immunol 6(6):593–599. Available from: http://www.nature.com/ni/journal/v6/n6/pdf/ni1201.pdf
39. Pawaria S, Messmer MN, Zhou YJ, Binder RJ (2011) A role for the heat shock protein-CD91 axis in the initiation of immune responses to tumors. Immunol Res 50(2–3):255–260
40. Triantafilou M, Triantafilou K (2004) Heat-shock protein 70 and heat-shock protein 90 associate with toll-like receptor 4 in response to bacterial lipopolysaccharide. Biochem Soc Trans 32(Pt 4):636–639. Available from: http://www.ncbi.nlm.nih.gov/pubmed/15270695\n http://www.biochemsoctrans.org/bst/032/0636/0320636.pdf
41. Zhou YJ, Messmer MN, Binder RJ (2015) NIH Public Access 2(3):217–228
42. Reyes-del Valle J, Chavez-Salinas S, Medina F, Del Angel RM (2005) Heat shock protein 90 and heat shock protein 70 are components of dengue virus receptor complex in human cells. J Virol 79(8):4557–4567. Available from: http://jvi.asm.org/content/79/8/4557.full.pdf+html
43. Wilson MR, Yerbury JJ, Poon S (2008) Extracellular chaperones and amyloids. In: Asea A, Brown I (eds) Heat shock proteins and the brain: implications for neurodegenerative diseases and neuroprotection. Springer Science, New York, pp 283–315

44. Csala M, Kereszturi É, Mandl J, Bánhegyi G (2012) The endoplasmic reticulum as the extracellular space inside the cell: role in protein folding and glycosylation. Antioxid Redox Signal 16(10):1100–1108. Available from: http://www.ncbi.nlm.nih.gov/pubmed/22149109
45. Sitia R, Braakman I (2003) Quality control in the endoplasmic reticulum protein factory. Nature 426:891–894
46. Garrido C, Gurbuxani S, Ravagnan L, Kroemer G (2001) Heat shock proteins: endogenous modulators of apoptotic cell death. Biochem Biophys Res Commun 286(3):433–442
47. Pratt B (1998) The hsp90-based chaperone system: involvement in signal transduction from a variety of hormone and growth factor receptors. Soc Exp Biol Med 217:420–434
48. Mayer MP, Bukau B (2005) Hsp70 chaperones: cellular functions and molecular mechanism. Cell Mol Life Sci 62(6):670–684
49. Taipale M, Krykbaev I, Koeva M, Kayatekin C, Westover D, Karras GI et al (2012) Quantitative analysis of Hsp90-client interactions reveals principles of substrate recognition. Cell 150(5):987–1001
50. Ullrich SJ, Robinson EA, Law LW, Willinghamt M, Appella E (1986) A mouse tumor-specific transplantation antigen is a heat shock related protein. Proc Natl Acad Sci 83:3121–3125
51. Sidera K, Samiotaki M, Yfanti E, Panayotou G, Patsavoudi E (2004) Involvement of cell surface HSP90 in cell migration reveals a novel role in the developing nervous system. J Biol Chem 279(44):45379–45388
52. Sims JD, McCready J, Jay DG (2011) Extracellular heat shock protein (Hsp)70 and Hsp90α assist in matrix metalloproteinase-2 activation and breast cancer cell migration and invasion. PLoS One 6(4):e18848. Available from: http://www.pubmedcentral.nih.gov/articlerender.fcgi?artid=3077417&tool=pmcentrez&rendertype=abstract. Cited 24 Jan 2014
53. Hunter MC, O'Hagan KL, Kenyon A, Dhanani KCH, Prinsloo E, Edkins AL (2014) Hsp90 binds directly to fibronectin (FN) and inhibition reduces the extracellular fibronectin matrix in breast cancer cells. PLoS One 9(1):e86842. Available from: http://dx.plos.org/10.1371/journal.pone.0086842. Cited 27 Jan 2014
54. Eustace BK, Jay DG (2004) Extracellular Roles for the molecular chaperone, Hsp90. Cell Cycle 3(9):1098–1100
55. Shin BK, Wang H, Yim AM, Le Naour F, Brichory F, Jang JH et al (2003) Global profiling of the cell surface proteome of cancer cells uncovers an abundance of proteins with chaperone function. J Biol Chem 278(9):7607–7616
56. McKeown-Longo PJ, Mosher DF (1985) Interaction of the 70,000-mol-wt amino-terminal fragment of fibronectin with the matrix-assembly receptor of fibroblasts. J Cell Biol 100(2):364–374. Available from: http://www.pubmedcentral.nih.gov/articlerender.fcgi?artid=2113439&tool=pmcentrez&rendertype=abstract
57. To WS, Midwood KS (2011) Plasma and cellular fibronectin: distinct and independent functions during tissue repair. Fibrogenesis Tissue Repair 4(1):21. Available from: http://www.pubmedcentral.nih.gov/articlerender.fcgi?artid=3182887&tool=pmcentrez&rendertype=abstract. Cited 28 Apr 2014
58. Sottile J, Hocking DC (2002) Fibronectin polymerization regulates the composition and stability of extracellular matrix fibrils and cell-matrix adhesions. Mol Biol Cell 13:3546–3559
59. Wierzbicka-Patynowski I, Schwarzbauer JE (2003) The ins and outs of fibronectin matrix assembly. J Cell Sci 116(Pt 16):3269–3276. Available from: http://www.ncbi.nlm.nih.gov/pubmed/12857786. Cited 5 Mar 2013
60. Ohashi T, Erickson HP (2009) Revisiting the mystery of fibronectin multimers: the fibronectin matrix is composed of fibronectin dimers cross-linked by non-covalent bonds. Matrix Biol 28(3):170–175. Available from: http://www.pubmedcentral.nih.gov/articlerender.fcgi?artid=2683204&tool=pmcentrez&rendertype=abstract. Cited 9 Nov 2014
61. Arap MA, Lahdenranta J, Mintz PJ, Hajitou A, Sarkis AS, Arap W et al (2004) Cell surface expression of the stress response chaperone GRP78 enables tumor targeting by circulating ligands. Cancer Cell 6(3):275–284

62. González-Ramos M, Calleros L, López-Ongil S, Raoch V, Griera M, Rodríguez-Puyol M et al (2013) HSP70 increases extracellular matrix production by human vascular smooth muscle through TGF-β1 up-regulation. Int J Biochem Cell Biol 45(2):232–242. Available from: http://www.ncbi.nlm.nih.gov/pubmed/23084979
63. Nagai N, Hosokawa M, Itohara S, Adachi E, Matsushita T, Hosokawa N et al (2000) Embryonic lethality of molecular chaperone Hsp47 knockout mice is associated with defects in collagen biosynthesis. J Cell Biol 150(6):1499–1505
64. Zhu J, Xiong G, Fu H, Evers BM, Zhou BP, Xu R (2015) Chaperone Hsp47 drives malignant growth and invasion by modulating an ECM gene network. Cancer Res 75(8):1580–1591
65. Taguchi T, Razzaque MS (2007) The collagen-specific molecular chaperone HSP47: is there a role in fibrosis? Trends Mol Med 13(2):45–53
66. Ravindran S, Gao Q, Ramachandran A, Blond S, Predescu A, George A (2011) Stress chaperone Grp78 functions in mineralized matrix formation. J Biol Chem 286(11):8729–8739
67. Li Z, Zhang L, Zhao Y, Li H, Xiao H, Fu R et al (2013) Cell-surface GRP78 facilitates colorectal cancer cell migration and invasion. Int J Biochem Cell Biol 45(5):987–994. Available from: http://www.ncbi.nlm.nih.gov/pubmed/23485528
68. Zheng J, Luo W, Tanzer ML (1998) Aggrecan synthesis and secretion: a paradigm for molecular and cellular coordination of multiglobular protein folding and intracellular trafficking a paradigm for molecular and cellular coordination of multiglobular protein. J Biol Chem 273(21):12999–13006
69. de Bock C, Lin Z, Mekkawy H, Byrne A, Wang Y (2010) Down-regulation of CacyBP is associated with poor prognosis and the effects on COX-2 expression in breast cancer. Int J Oncol 36:1155–1163
70. Lin Y, Peng N, Zhuang H, Zhang D, Wang Y, Hua ZC (2014) Heat shock proteins HSP70 and MRJ cooperatively regulate cell adhesion and migration through urokinase receptor. BMC Cancer 14:639. Available from: http://www.ncbi.nlm.nih.gov/pubmed/25175595
71. Bartl MM, Luckenbach T, Bergner O, Ullrich O, Koch-Brandt C (2001) Multiple receptors mediate apoJ-dependent clearance of cellular debris into nonprofessional phagocytes. Exp Cell Res 271(1):130–141. Available from: http://www.ncbi.nlm.nih.gov/pubmed/11697889
72. Zlokovic BV, Martel CL, Matsubara E, McComb JG, Zheng G, McCluskey RT et al (1996) Glycoprotein 330/megalin: probable role in receptor-mediated transport of apolipoprotein J alone and in a complex with Alzheimer disease amyloid beta at the blood-brain and blood-cerebrospinal fluid barriers. Proc Natl Acad Sci U S A 93(9):4229–4234
73. Tsutsumi S, Scroggins B, Koga F, Lee M-J, Trepel J, Felts S et al (2008) A small molecule cell-impermeant Hsp90 antagonist inhibits tumor cell motility and invasion. Oncogene 27(17):2478–2487. Available from: http://www.ncbi.nlm.nih.gov/pubmed/17968312. Cited 21 Apr 2013
74. Wozniak MA, Modzelewska K, Kwong L, Keely PJ (2004) Focal adhesion regulation of cell behavior. Biochim Biophys Acta 1692:103–119
75. Haas TL (2005) Endothelial cell regulation of matrix metalloproteinases. Can J Physiol Pharmacol 83(1):1–7. Available from: http://www.ncbi.nlm.nih.gov/pubmed/15759044
76. McCready J, Wong DS, Burlison JA, Ying W, Jay DG (2014) An impermeant ganetespib analog inhibits extracellular Hsp90-mediated cancer cell migration that involves lysyl oxidase 2-like protein. Cancers 6(2):1031–1046. Available from: http://www.pubmedcentral.nih.gov/articlerender.fcgi?artid=4074815&tool=pmcentrez&rendertype=abstract. Cited 20 Mar 2015
77. Anelli T, Sitia R (2008) Protein quality control in the early secretory pathway. EMBO J 27(2):315–327. Available from: http://www.pubmedcentral.nih.gov/articlerender.fcgi?artid=2234347&tool=pmcentrez&rendertype=abstract
78. Hebert C, Norris K, Della CR, Reynolds M (1999) Cell surface colligin/Hsp47 associates with tetraspanin protein CD9 in epidermoid carcinoma cell lines. J Cell Biochem 73(3):248–258
79. Ni M, Lee AS (2007) ER chaperones in mammalian development and human diseases. FEBS Lett 581(19):3641–3651
80. Staron M, Yang Y, Liu B, Li J, Shen Y, Zuniga-Pflucker JC et al (2010) gp96, an endoplasmic reticulum master chaperone for integrins and Toll-like receptors, selectively regulates early T and B lymphopoiesis. Blood 115(12):2380–2390

81. Luo B, Lam BS, Lee SH, Wey S, Zhou H, Wang M et al (2011) The endoplasmic reticulum chaperone protein GRP94 is required for maintaining hematopoietic stem cell interactions with the adult bone marrow niche. PLoS One 6(5):e20364
82. Oldberg A, Antonsson P, Lindblom K, Heinegård D (1992) COMP (cartilage oligomeric matrix protein) is structurally related to the thrombospondins. J Biol Chem 267(31):22346–22350. Available from: http://www.ncbi.nlm.nih.gov/pubmed/1429587
83. Hecht JT, Nelson LD, Crowder E, Wang Y, Elder FF, Harrison WR, Francomano CA, Prange CK, Lennon GG, Deere M, Lawler J (1995) Mutations in exon 17B of cartilage oligomeric matrix protein (COMP) cause pseudoachondroplasia. Nat Genet 10(3):325–329
84. Horton WA, Hecht JT (2001) Skeletal dysplasias, chondrodysplasias: disorders of cartilage matrix proteins chondrodysplasias. In: Royce MP, Steinmann B (eds) Connective tissue and its heritable disorders: molecular, genetic and medical aspects, 2nd edn. John Wiley and Sons Inc., New York, pp 909–937
85. Hecht JT, Hayes E, Snuggs M, Decker G, Montufar-Solis D, Doege K et al (2001) Calreticulin, PDI, Grp94 and BiP chaperone proteins are associated with retained COMP in pseudoachondroplasia chondrocytes. Matrix Biol 20(4):251–262
86. Hecht JT, Deere M, Putnam E, Cole W, Vertel B, Chen H et al (1998) Characterization of cartilage oligomeric matrix protein (COMP) in human normal and pseudoachondroplasia musculoskeletal tissues. Matrix Biol 17(4):269–278
87. Richter K, Buchner J (2011) Closing in on the Hsp90 chaperone-client relationship. Structure 19(4):445–446. https://doi.org/10.1016/j.str.2011.03.007
88. Doran AC, Meller N, Mcnamara CA (2009) The role of smooth muscle cells in the initiation and early progression of atherosclerosis. Arterioscler Thromb Vasc Biol 28(5):812–819
89. Kanwar RK, Kanwar JR, Wang D, Ormrod DJ, Krissansen GW (2001) Temporal expression of heat shock proteins 60 and 70 at lesion-prone sites during atherogenesis in ApoE-deficient mice. Arterioscler Thromb Vasc Biol 21(12):1991–1997. Available from: http://www.ncbi.nlm.nih.gov/pubmed/11742875
90. Philips N, Keller T, Gonzalez S (2004) TGF beta-like regulation of matrix metalloproteinases by anti-transforming growth factor-beta, and anti-transforming growth factor-beta 1 antibodies in dermal fibroblasts: implications for wound healing. Wound Repair Regen 12(1):53–59. Available from: http://www.ncbi.nlm.nih.gov/pubmed/14974965
91. Ruiz-Ortega M, Rodríguez-Vita J, Sanchez-Lopez E, Carvajal G, Egido J (2007) TGF-β signaling in vascular fibrosis. Cardiovasc Res 74(2):196–206
92. Ong VH, Carulli MT, Xu S, Khan K, Lindahl G, Abraham DJ et al (2009) Cross-talk between MCP-3 and TGFbeta promotes fibroblast collagen biosynthesis. Exp Cell Res 315(2):151–161. https://doi.org/10.1016/j.yexcr.2008.11.001
93. Asea A, Kabingu E, Stevenson MA, Calderwood SK (2000) HSP70 peptide-bearing and peptide-negative preparations act as chaperokines. Cell Stress Chaperones 5(5):425–431. Available from: http://www.pubmedcentral.nih.gov/articlerender.fcgi?artid=312872&tool=pmcentrez&rendertype=abstract
94. Delpino A, Castelli M (2002) The 78 kDa glucose-regulated protein (GRP78/BIP) is expressed on the cell membrane, is released into cell culture medium and is also present in human peripheral circulation. Biosci Rep 22(3–4):407–420
95. Hayakawa K, Hiramatsu N, Okamura M, Yamazaki H, Nakajima S, Yao J et al (2009) Acquisition of anergy to proinflammatory cytokines in nonimmune cells through endoplasmic reticulum stress response: a mechanism for subsidence of inflammation. J Immunol 182(2):1182–1191. Available from: http://www.ncbi.nlm.nih.gov/pubmed/19124762
96. Misra K, Deedwania R, Pizzo SV (2005) Binding of activated a2-macroglobulin to its cell surface receptor Grp78 in 1-LN prostate cancer cells regulates PAK-2 dependent activation of LIMK. J Biol Chem 280(28):26278–26286
97. Freyria AM, Ronziere MC, Boutillon MM, Herbage D (1995) Effect of retinoic acid on protein synthesis by foetal bovine chondrocytes in high-density culture: down-regulation of the glucose-regulated protein, GRP-78, and type II collagen. Biochem J 305:391–396. Available from: http://www.ncbi.nlm.nih.gov/pubmed/7832751

98. Ridley AJ, Schwartz MA, Burridge K, Firtel RA, Ginsberg MH, Borisy G et al (2003) Cell migration: integrating signals from front to back. Science 302:1704–1710
99. Kampinga HH, Craig EA (2010) The HSP70 chaperone machinery: J proteins as drivers of functional specificity. Nat Rev 11(8):579–592. Available from: http://www.pubmedcentral.nih.gov/articlerender.fcgi?artid=3003299&tool=pmcentrez&rendertype=abstract
100. Mitra A, Fillmore RA, Metge BJ, Rajesh M, Xi Y, King J et al (2008) Large isoform of MRJ (DNAJB6) reduces malignant activity of breast cancer. Breast Cancer Res 10(2):1–13
101. Sterrenberg JN, Blatch GL, Edkins AL (2011) Human DNAJ in cancer and stem cells. Cancer Lett 312(2):129–142. https://doi.org/10.1016/j.canlet.2011.08.019
102. Qiu XB, Shao YM, Miao S, Wang L (2006) The diversity of the DnaJ/Hsp40 family, the crucial partners for Hsp70 chaperones. Cell Mol Life Sci 63(22):2560–2570
103. Cheetham ME, Caplan AJ (1998) Structure, function and evolution of DnaJ: conservation and adaptation of chaperone function. Cell Stress Chaperones 3:28–36
104. Cyr MD, Ramos HC (2015) Specification of Hsp70 function by type I and type II Hsp40. In: Blatch LG, Edkins LA (eds) The networking of chaperones by co-chaperones. Springer, New York, pp 91–102
105. Rose JM, Novoselov SS, Robinson PA, Cheetham ME (2011) Molecular chaperone-mediated rescue of mitophagy by a Parkin RING1 domain mutant. Hum Mol Genet 20(1):16–27
106. Blasi F, Carmeliet P (2002) uPAR: a versatile signalling orchestrator. Nat Rev Mol Cell Biol 3(12):932–943
107. Madsen CD, Ferraris SM, Andolfo A, Cunningham O, Sidenius N (2007) uPAR-induced cell adhesion and migration: vitronectin provides the key. J Cell Biol 177(5):927–939
108. Kenny HA, Kaur S, Coussens LM, Lengyel E (2008) The initial steps of ovarian cancer cell metastasis are mediated by MMP-2 cleavage of vitronectin and fibronectin. J Clin Invest 118(4):1367–1379
109. Wei Y, Lukashev M, Simon DI, Bodary SC, Rosenberg S, Doyle MV et al (1996) All use subject to JSTOR terms and conditions regulation of integrin function by the urokinase receptor. Science 273(5281):1551–1555
110. Aguirre-Ghiso JA, Liu D, Mignatti A, Kovalski K, Ossowski L (2001) Urokinase receptor and fibronectin regulate the ERK MAPK to p38 MAPK activity ratios that determine carcinoma cell proliferation or dormancy in vivo. Mol Biol Cell 12:863–879
111. Waltz DA, Natkin LR, Fujita RM, Wei Y, Chapman HA (1997) Plasmin and plasminogen activator inhibitor type 1 promote cellular motility by regulating the interaction between the urokinase receptor and vitronectin. J Clin Invest 100(1):58–67
112. Calderwood SK, Khaleque MA, Sawyer DB, Ciocca DR (2006) Heat shock proteins in cancer: chaperones of tumorigenesis. Trends Biochem Sci 31(3):164–172
113. Das S, Harris LG, Metge BJ, Liu S, Riker AI, Samant RS et al (2009) The hedgehog pathway transcription factor GLI1 promotes malignant behavior of cancer cells by up-regulating osteopontin. J Biol Chem 284(34):22888–22897
114. Mitra A, Menezes ME, Pannell L, Mulekar M, Honkanen R, Shevde L et al (2012) DNAJB6 chaperones PP2A mediated dephosphorylation of GSK3β to downregulate β-catenin transcription target, osteopontin. Oncogene 31(41):4472–4483
115. Davids W, El-Thaher HS, Nakai A, Nagata K, Miller D (1995) Modelling the three-dimensional structure of serpin/molecular chaperone hsp47. Bioorg Chem 23:427–438
116. Kurkinen M, Taylor A, Garrels JI, Hogan BLM (1984) Cell surface-associated proteins which bind native type IV collagen or gelatin. J Biol Chem 259(9):5915–5922
117. Nagata K (1996) Hsp47: a collagen-specific molecular chaperone. TIBS Rev 21:23–26
118. Kadler KE, Holmes DF, Trotter JA, Chapman JA (1996) Collagen fibril formation. J Biochem 316(Pt 1):1–11
119. Masago Y, Hosoya A, Kawasaki K, Kawano S, Nasu A, Toguchida J et al (2012) The molecular chaperone Hsp47 is essential for cartilage and endochondral bone formation. J Cell Sci 125(Pt 5):1118–1128. Available from: http://www.ncbi.nlm.nih.gov/pubmed/22492985

120. Ishida Y, Nagata K (2011) Hsp47 as a collagen-specific molecular chaperone. Methods Enzymol 499:167–182
121. Wilson MR, Easterbrook-Smith SB (2000) Clusterin is a secreted mammalian chaperone. Trends Biochem Sci 25(3):95–98
122. DJS H, Bruns RR (1983) On the state of aggregation of newly secreted procollagen. Proc Natl Acad Sci U S A 80:388–392
123. Lamandé SR, Bateman JF (1999) Procollagen folding and assembly: the role of endoplasmic reticulum enzymes and molecular chaperones. Semin Cell Dev Biol 10(5):455–464
124. Martinek N, Shahab J, Sodek J, Ringuette M (2007) Is an evolutionarily conserved collagen chaperone? J Dent Res 86(4):296–305
125. Hosokawa N, Hohenadl C, Satoh M, Kühn K, Nagata K (1998) HSP47, a collagen-specific molecular chaperone, delays the secretion of type III procollagen transfected in human embryonic kidney cell line 293: a possible role for HSP47 in collagen modification. J Biochem 124(3):654–662. Available from: http://www.ncbi.nlm.nih.gov/pubmed/9722680
126. Zhang X, Yang J-J, Kim S, Kim K-Y, Ahn S, Yang S (2010) An 8-gene signature, including methylated and down-regulated glutathione peroxidase 3, of gastric cancer. Int J Oncol 36:405–414
127. Curran CS, Keely PJ (2013) Breast tumor and stromal cell responses to TGF-β and hypoxia in matrix deposition. Matrix Biol 32(2):95–105. https://doi.org/10.1016/j.matbio.2012.11.016
128. Chou J, Lin JH, Brenot A, Kim J, Provot S, Werb Z (2013) GATA3 suppresses metastasis and modulates the tumour microenvironment by regulating microRNA-29b expression. Nat Cell Biol 15(2):201–213. Available from: http://www.pubmedcentral.nih.gov/articlerender.fcgi?artid=3660859&tool=pmcentrez&rendertype=abstract
129. Sauk JJ, Nikitakis N, Siavash H (2005) Hsp47 a novel collagen binding serpin chaperone, autoantigen and therapeutic target. Front Biosci 10:107–118. Available from: http://www.ncbi.nlm.nih.gov/pubmed/15574354
130. Sato Y, Murase K, Kato J, Kobune M, Sato T, Kawano Y et al (2008) Resolution of liver cirrhosis using vitamin A-coupled liposomes to deliver siRNA against a collagen-specific chaperone. Nat Biotechnol 26(4):431–442
131. Jakob U, Gaestel M, Engel K, Buchner J (1993) Small heat shock proteins are molecular chaperones. J Biol Chem 268(3):1517–1520
132. Watanabe H, Yamada Y, Kimata K (1998) Roles of aggrecan, a large chondroitin sulfate proteoglycan, in cartilage structure and function. J Biochem 124(4):687–693
133. Tiffee JC, Griffin JP, Cooper LF (2000) Immunolocalization of stress proteins and extracellular matrix proteins in the rat tibia. Tissue Cell 32(2):141–147. Available from: http://www.ncbi.nlm.nih.gov/pubmed/10855699
134. Bailey RW, Aronow B, Harmony JAK, Griswold MD (2002) Heat shock-initiated apoptosis is accelerated and removal of damaged cells is delayed in the testis of clusterin/ApoJ knock-out mice. Biol Reprod 66(4):1042–1053
135. Michel D, Chatelain G, North S, Brun G (1997) Stress-induced transcription of the clusterin/apoJ gene. Biochem J 328:45–50. Available from: http://www.pubmedcentral.nih.gov/articlerender.fcgi?artid=1218885&tool=pmcentrez&rendertype=abstract
136. Humphreys DT, Carver JA, Easterbrook-Smith SB, Wilson MR (1999) Clusterin has chaperone-like activity similar to that of small heat shock proteins. J Biol Chem 274(11):6875–6881
137. Chaari A, Hoarau-Vechot J, Ladjimi M (2013) Applying chaperones to protein-misfolding disorders: molecular chaperones against alpha-synuclein in Parkinson's disease. Int J Biol Macromol 60:196–205. https://doi.org/10.1016/j.ijbiomac.2013.05.032
138. Kim WS, Kågedal K, Halliday GM (2014) Alpha-synuclein biology in Lewy body diseases. Alzheimers Res Ther 6(5):73. Available from: http://alzres.com/content/6/5/73
139. French K, Yerbury JJ, Wilson MR (2008) Protease activation of α2-macroglobulin modulates a chaperone-like action with broad specificity. Biochemistry 47(4):1176–1185

140. Narita M, Holtzman DM, Schwartz AL, Bu G (1997) Alpha2-macroglobulin complexes with and mediates the endocytosis of Beta-amyloid peptide via cell surface low-density lipoprotein receptor-related protein. J Neurochem 69(5):1904–1911
141. Shibata M, Yamada S, Ram Kumar S, Calero M, Bading J, Frangione B et al (2000) Clearance of Alzheimer's amyloid beta1-40 peptide from brain by LDL receptor-related protein-1 at the blood-brain barrier. J Clin Invest 106(12):1489–1499
142. Lane TF, Sage EH (1994) The biology of SPARC, a protein that modulates cell-matrix interactions. FASEB J 8(2):163–173
143. Termine JD, Belcourt AB, Conn KM, Kleinman HK (1981) Mineral and collagen-binding proteins of fetal calf bone. J Biol Chem 256(20):10403–10408
144. Lane TF, Iruela-Arispe ML, Sage EH (1992) Regulation of gene expression by SPARC during angiogenesis in vitro. Changes in fibronectin, thrombospondin-1, and plasminogen activator inhibitor-1. J Biol Chem 267(23):16736–16745
145. Tremble PM, Lane TF, Sage EH, Werb Z (1993) SPARC, a secreted protein associated with morphogenesis and tissue remodeling, induces expression of metalloproteinases in fibroblasts through a novel extracellular matrix-dependent pathway. J Cell Biol 121(6):1433–1444
146. Ingham KC, Brew SA, Migliorini M (2002) Type I collagen contains at least 14 cryptic fibronectin binding sites of similar affinity. Arch Biochem Biophys 407(2):217–223
147. Pearl LH, Prodromou C, Workman P (2008) The Hsp90 molecular chaperone: an open and shut case for treatment. Biochem J 410(3):439–453. Available from: http://www.ncbi.nlm.nih.gov/pubmed/18290764. Cited 25 May 2013
148. Pratt WB, Toft DO (2003) Regulation of signaling protein function and trafficking by the Hsp90/Hsp70-based chaperone machinery. Exp Biol Med 228:111–133
149. Corriden R, Insel AP (2011) Basal release of ATP: an autocrine-paracrine mechanism for cell regulation. Sci Signal 3(104):1–57. Available from: http://www.pubmedcentral.nih.gov/articlerender.fcgi?artid=3146065&tool=pmcentrez&rendertype=abstract
150. Fitz JG (2007) Regulation of cellular AtP release. Trans Am Clin Clim Assoc 118:199–208. Available from: http://www.pubmedcentral.nih.gov/articlerender.fcgi?artid=1863605&tool=pmcentrez&rendertype=abstract
151. Orriss IR, Key ML, Hajjawi MOR, Arnett TR (2013) Extracellular ATP released by osteoblasts is a key local inhibitor of bone mineralisation. PLoS One 8(7):1–13
152. Poon S, Easterbrook-Smith SB, Rybchyn MS, Carver JA, Wilson MR (2000) Clusterin is an ATP – independent chaperone with very broad substrate specificity that stabilizes stressed proteins in a folding-competent state. Biochemistry 39(51):15953–15960
153. Li W, Sahu D, Tsen F (2011) Secreted heat shock protein-90 (Hsp90) in wound healing and cancer. Biochim Biophys Acta 1823(3):730–741. Available from: http://www.ncbi.nlm.nih.gov/entrez/query.fcgi?cmd=Retrieve&db=PubMed&dopt=Citation&list_uids=21982864. Cited 6 Mar 2013
154. Khalil AA, Kabapy NF, Deraz SF, Smith C (2011) Heat shock proteins in oncology: diagnostic biomarkers or therapeutic targets? Biochim Biophys Acta 1816(2):89–104. Available from: http://www.ncbi.nlm.nih.gov/pubmed/21605630. Cited 28 Oct 2013
155. Leavesley DI, Kashyap AS, Croll T, Sivaramakrishnan M, Shokoohmand A, Hollier BG et al (2013) Vitronectin–master controller or micromanager? IUBMB Life [Internet] 65(10): 807–818. Available from: http://www.ncbi.nlm.nih.gov/pubmed/24030926
156. Zhu J, Xiong G, Trinkle C, Xu R (2014) Integrated extracellular matrix signaling in mammary gland development and breast cancer progression. Histol Histopathol 29(9):1083–92
157. Preissner, KT, Reuning U (2011) Vitronectin in vascular context: facets of a multitalented matricellular protein. Seminars in thrombosis and hemostasis, pp 408–24
158. Halper J, Kjaer M (2013) No TitleBasic components of connective tissues and extracellular matrix: elastin, fibrillin, fibulins, fibrinogen, fibronectin, laminin, tenascins and thrombospondins. In: Halper J (ed) Progress in heritable soft connective tissue diseases. Springer, pp 31–47

159. Bellahcène A, Castronovo V, Ogbureke KUE, Fisher LW, Fedarko NS (2008) Small integrin-binding ligand N-linked glycoproteins (SIBLINGs): multifunctional proteins in cancer. Nat Rev Cancer [Internet] 8(3):212–26. Available from: http://dx.doi.org/10.1038/nrc2345
160. Narayanan K, Ramachandran A, Hao J, He G, Park KW, Cho M et al (2003) Dual functional roles of dentin matrix protein 1. Implications in biomineralization and gene transcription by activation of intracellular Ca2+ store. J Biol Chem 278(19):17500–17508
161. Chlenski A, Guerrero LJ, Salwen HR, Yang Q, Tian Y, la Madrid A et al (2011) Secreted protein acidic and rich in cysteine is a matrix scavenger chaperone. PLoS One 6(9):e23880
162. Adams JC, Lawler J (2011) The Thrombospondins. Cold Spring Harbor Perspectives in Biology [Internet]. Available from: http://www.pubmedcentral.nih.gov/articlerender.fcgi?artid=3179333&tool=pmcentrez&rendertype=abstract
163. Sodek J, Ganss B, McKee D (2000) Osteopontin. Crit Rev Oral Biol Med 1(3):279–303
164. Oskarsson T, Acharyya S, Zhang XH-F, Vanharanta S, Tavazoie SF, Morris PG et al (2011) Breast cancer cells produce tenascin C as a metastatic niche component to colonize the lungs. Nat Med [Internet]. 17(7):867–74. Available from: http://www.pubmedcentral.nih.gov/articlerender.fcgi?artid=4020577&tool=pmcentrez&rendertype=abstract

# Chapter 7
# Roles, Mechanisms, and Opportunities of Heat Shock Protein gp96/grp94 in Infections and Inflammation-Associated Malignancies

**Songdong Meng and Zihai Li**

**Abstract** Heat shock proteins (HSP) gp96 (grp94) play an important role in modulating innate and T cell immunity via interaction with toll-like receptors (TLRs) and chaperoning antigenic peptides for antigen presentation to MHC molecules. These immunological properties of gp96 have inspired development of gp96-based prophylactic and therapeutic vaccines against various pathogens, including influenza virus, human papillomavirus, *Mycobacterium tuberculosis*, hepatitis B virus, and herpes simplex virus in mice models. Besides the already known underlying counterback mechanisms, the intrinsic characteristic of gp96 that simultaneously induce both effector T cell response and regulatory T cells (Tregs) may account for the modest efficiency of gp96-based immunotherapy against chronic infections and cancer. There is thus a strong need for identifying novel combination strategies (e.g., Treg inhibition, and immune checkpoint targeting) for designing a more effective gp96-based vaccine against pathogen infections. In addition, targeting cell membrane gp96 might provide a novel therapeutic approach as certain pathogens induce translocation of endoplasmic reticulum-resided gp96 to cell surface. Placenta-derived gp96 has the ability to initiate antitumor T-cell immunity via association with multiple embryo-cancer antigens against chronic infection-associated cancers. Further understanding of the placental gp96 associated-carcinoembryonic antigen repertoires that orchestrate immune defense networks against pathogen-induced cancer formation may allow ample opportunities to provide an effective strategy in cancer prevention and therapy.

S. Meng (✉)
Key Laboratory of Pathogenic Microbiology and Immunology, Institute of Microbiology, Chinese Academy of Sciences, Beijing, China
e-mail: mengsd@im.ac.cn

Z. Li
Hollings Cancer Center, Department of Microbiology and Immunology, Medical University of South Carolina, Charleston, SC, USA
e-mail: zihai@musc.edu

 
R. J. Binder, P. K. Srivastava (eds.), *Heat Shock Proteins in the Immune System*,
https://doi.org/10.1007/978-3-319-69042-1_7

**Keywords** Heat shock protein · gp96 · T cell immunity · Toll-like receptors · Treg · Vaccine · Adjuvant · Immunotherapy

## Abbreviations

| | |
|---|---|
| ALT | Alanine aminotransaminase |
| APC | Antigen-presenting cell |
| BCG | Bacillus Calmette-Guerin |
| Con A | Concanavalin A |
| CTL | Cytotoxic T lymphocyte |
| CTLA4 | Cytotoxic T lymphocyte-associated antigen-4 |
| DC | Dendritic cell |
| DENA | Diethylnitrosamine |
| Foxp3 | Forkhead/winged helix transcription factor |
| hbcag | Hepatitis B core antigen |
| hbsag | Hepatitis B surface antigen |
| HBV | Hepatitis B virus |
| HCC | Hepatocellular carcinoma |
| HCV | Hepatitis C virus |
| HPV | Human papillomavirus |
| HSP | Heat shock protein |
| HSV | Herpes simplex virus |
| IL | Interleukin |
| LPS | Lipopolysaccharide |
| mab | Monoclonal antibody |
| MHC | Major histocompatibility complex |
| MDP | Muramyl dipeptide |
| MAPK | Mitogen-activated protein kinase |
| NLRP3 | NLR family, pyrin domain containing 3 |
| PD-1 | Programmed death-1 |
| PD-L1 | Programmed death 1 ligand 1 |
| TAP | Transporter associated with antigen processing |
| TB | Tuberculosis |
| TLR | Toll like receptor |
| TNF | Tumor necrosis factor |

## 7.1 Introduction

Both humoral and cellular immunity play critical roles in the control of pathogen infections. In general, humoral immunity plays major roles in prophylactic activity by neutralizing antibody against invading pathogens, whereas cellular immunity is believed to play key roles for control and clearance of intracellular pathogens (e.g., virus and *Mycobacterium*). The elimination of intracellular pathogens is mainly mediated by robust antigen-specific CD8+ T cell (CTL) and helper T cell responses for acute and self-limited infections, while chronic infection is characterized by weak T-cell responses [1–3].

Heat shock protein (HSP) gp96 (glucose-regulated protein 94, GRP94) is an endoplasmic reticulum (ER)-resident chaperone that has the unique capability to associate with antigenic peptides. Both rodent models and clinical trials have demonstrated that gp96 purified from tumors or complexed with viral antigens in vitro elicits antitumor effects or antigen-specific CTL immunity against tumors and viruses [4–6]. The immunogenicity of gp96 is attributed to its ability to activate both the innate and adaptive immune responses, as summarized below. First, together with HSP70 and HSP90 in the cytosol, gp96, TAP (transporter associated with antigen processing) molecules, and calreticulin in the ER are thought to constitute a relay line for antigenic peptide transfer from the cytosol to MHC class I molecules in a concerted and regulated manner [7]. Under intradermal or subcutaneous immunization, gp96-antigen complexes access the draining lymph node and enter professional antigen-presenting cells (APCs) such as dendritic cells (DCs) through gp96 receptor CD91-mediated or/and scavenger receptor-mediated mechanisms. The associated antigens are eventually presented to MHC I and II molecules [8–10]. This leads to activation of antigen-specific CD8+ and CD4+ T cell responses. Second, gp96 itself binds to and acts as a master chaperone for Toll-like receptors (TLRs) (e.g., TLR2, TLR4, and TLR9) on APCs, stimulating proinflammatory and Th1-type cytokine (TNF-α, IL-1β, and IL-12) secretion [11, 12]. Third, gp96 also interacts with CD91, which leads to CD91 phosphorylation and activation of NF-κB and p38 MAPK. This allows for the maturation of APCs, releasing cytokines, and priming of T-helper (Th) cells [13].

Currently, two major types of adjuvants are used in human vaccines against viral and bacterial infections, i.e., aluminum salts-based adjuvants and oil-in-water emulsions. Aluminum salts are mainly used to improve humoral immune responses and the polarized Th2 cell response in a vaccine, likely via NLRP3 inflammasome activation or the release of the endogenous danger signal, uric acid [14]. However, their capability to stimulate a cellular immune response is rather limited [15]. The oil-in-water emulsion-based adjuvants (such as AS03 (GlaxoSmithKline (GSK), MF59 (Novartis), and AF03 (Sanofi Pasteur)) are currently the most widely used component for vaccines and promote a mixed Th1 and Th2 cell response, likely by activating DCs or increasing antigen uptake [16]. However, the currently used adjuvant vaccines in general could not effectively induce potent and cross-protective T cell immunity against pathogen infections, although a broader antibody response is observed. Given the critical role of T cell-mediated immune responses in protection against intracellular pathogen infections, especially in control of established infections, it is urgently necessary to explore new adjuvants for prophylactic and therapeutic vaccines that can augment T cell responses.

## 7.2 HSPs-Mediated Balance Between Regulatory and Responder T Cells

### *7.2.1 Enhanced Treg Function by gp96*

In both mice and humans, the Treg population is identified by high expression of IL-2R a chain (CD25). Forkhead/winged helix transcription factor (Foxp3) is also expressed by and required for development and function of Tregs. The widespread distribution of Tregs as a key checkpoint in both lymphoid and nonlymphoid tissues, and the selective recruitment of Tregs to different tissue sites control tissue inflammation in autoimmunity, infection, and cancer development [17]. CD4+CD25+Foxp3+ Treg suppresses the activation, proliferation, and effector functions of many cell types including CD4+ and CD8+ T cells, playing an important role in the maintenance of immunologic tolerance to self-antigen (Ag), as well as pathogens and tumors.

HSP gp96 and its N-terminal fragment were found to have the ability to augment CTL responses against pathogens such as hepatitis B virus (HBV). However, administration of higher amounts of gp96 was detrimental to the adjuvant effect and decreased the capability of mice to generate an immune T-cell response [18]. Titration of gp96 dose (0, 0.5, 5, 10, 20, 50, 100, and 200 μg/mice) demonstrated that immunization with 10–20 μg/mice of gp96 induced the highest CTL response in mice, which decreased dramatically when the immunization dose increased to 50–100 μg. Other studies involving gp96 as adjuvant have also indicated that high-dose gp96 administration can downregulate inflammatory events and antitumor effect. This result has been attributed to the activation of Treg or myeloid suppressor cells. Indeed, immunization with gp96 with HBV antigens simultaneously stimulated both antigen-specific CTL and Treg activity. Activation of CTL at low dose of gp96 was far more pronounced than Treg activation. Treg population and suppression increased with elevated gp96 dose, eventually abrogating the T-cell response induced by immunization. In addition, low dose cyclophosphamide treatment could restore the T-cell responses loss after high-dose gp96 adjuvant injection by suppression of Treg activation. These studies reveal that gp96-induced immune response appears to be the reflection of the overall effects of CTL and Treg [19].

### *7.2.2 Mechanisms of gp96-Mediated Treg Activation*

In murine genetic models that delete gp96 in a Treg lineage-specific fashion, gp96 is shown to be essential for Treg maintenance and function, as loss of gp96 resulted in instability of the Treg lineage and impairment of their suppressive functions. Tregs without gp96 were unable to maintain Foxp3 expression levels, resulting in systemic accumulation of pathogenic IFN-γ-producing and IL-17-producing T cells. Mechanistic study showed that gp96 plays essential roles in maintaining

TGF-β bioavailability and Treg function by chaperoning both GARP and integrins [20]. As Tregs are one of the major barriers impeding antipathogen and antitumor immune responses, suppression of Tregs by blocking gp96 would reverse immune tolerance and thus promote pathogen and cancer immunity [21].

TLR2 and TLR4 ligands and agonists modulate Treg proliferation, survival, and function by directly acting on the Treg population [22], and gp96 interacts with TLRs and activates innate immunity [11]. HSP gp96 directly binds to Tregs via interaction with TLR2 and TLR4, thus activating the NF-κB pathway, which promotes Foxp3, IL-10, and TGF-β1 expression. As Foxp3 and the suppressive cytokines IL-10 and TGF-β play key roles in Treg function, gp96 may modulate Tregs function via direct interaction with TLR2 and TLR4 on the surface of Tregs [23].

The activation of Tregs requires high-dose gp96 stimulation, whereas low-dose gp96 could induce effector T cells. This may be due to the limited access of gp96 to Tregs, which only account for ~3% of total splenocytes. Another explanation is that the anergy state of Tregs in a normal host requires a high amount of gp96 binding to its surface to activate downstream signaling.

### *7.2.3 Harness of the Two-Edged Sword of gp96 Activity*

Immunization with high-dose gp96 induces Treg activities. Thus, dose-dependent manipulation of Treg cell activity may have a therapeutic potential to restrain immune hyperactivation in autoimmunity, inflammation, and allograft rejection. As hepatic T lymphocytes and NK cells-mediated inflammation are involved in the pathogenesis of HBV-induced chronic liver diseases, Tregs play a key role in intrahepatic immune regulation. Treg frequency has been shown to be inversely correlated with immune-mediated liver injury and pathogenesis of HBV-associated fibrosis progression and liver failure [24]. In concanavalin A (Con A)- and anti-CD137-induced severe liver hepatitis mouse models, high-dose gp96 immunization elicited rapid and long-lasting protection of mice against liver injury, as evidenced by decreased number of parameters including alanine aminotransaminase (ALT) levels, hepatic necrosis, serum proinflammatory cytokines (IFN-γ, TNF-α, and IL-6), and number of IFN-γ+CD4+ and IFN-γ+CD8+ T cells in the spleen and liver. In contrast, CD4+CD25+Foxp3+ Treg frequency and suppressive function were both increased by high-dose gp96 stimulation, and the protective effect of gp96 could be generated by adoptive transfer of Treg cells from gp96-immunized mice [23]. Thus, high-dose gp96-based therapy could be developed against immune-mediated liver destruction in T-cell-mediated immune hyperactivation syndromes.

The induction of Treg activities by gp96 may also explain why vaccinations with autologous tumor-derived gp96 or complexes of gp96-pathogen-derived antigens have generated relatively modest antitumor activities in both rodents and clinical trials [25]. Notably, the blockade of Treg by a monoclonal antibody or low dose cyclophosphamide significantly increased gp96-HBV peptide complex-mediated anti-HBV CTL responses and synergistically enhanced gp96 tumor vaccine-induced antitumor immunity in mouse models [26]. As there are already drugs in clinical trials targeting Treg [27, 28], combination of gp96 vaccine with Treg inactivation is a promising strategy for breaking tumor immune evasion and deserves further evaluation for the treatment of chronic infection and cancer. Clearly, investigation on the mechanisms of Treg activation/inactivation during gp96 immunization is important for developing better combination strategies in the future.

## 7.3 HSP-Based Prophylactic Vaccines Against Pathogens

### *7.3.1 HSP-Adjuvanted Influenza Vaccine*

The emergence of pandemic influenza strains that have acquired interspecies transmission to human (e.g., the 2009 H1N1 swine virus epidemic worldwide or the more recent surge of the H7N9 avian virus in China) posed a great threat to public health. The vaccine-induced protective neutralizing antibodies (Abs) that target the outer hemagglutinin (HA) and neuraminidase (NA) proteins of influenza viruses are highly strain specific, and thus, the current vaccines against seasonal influenza strains are ineffective for different strains due to the variability of influenza A virus. Current split-virion vaccines with or without aluminum adjuvant induce a highly humoral immune response but fail to induce cytotoxic T-lymphocyte immunity, which plays a utilitarian role in inducing cross-protection against various subtypes of influenza virus as T cell epitopes are generally derived from conserved internal components of the virus [29–32].

Because conventional inactivated vaccines do not effectively induce CTLs, two current strategies are used to increase the magnitude of cellular immunity to vaccination and elicit robust cross-protection. One is the use of conserved viral antigens (e.g., NP and M2) or T cell epitopes to provide cross-protective immunity and develop universal influenza vaccines [31, 33]. The other is to incorporate mucosal adjuvants (e.g., chitosan, TLR3-specificdouble-stranded RNA oligonucleotide, or polyI:polyC) into the vaccines for intranasal immunization [34, 35]. It is also attractive to incorporate T cell adjuvants into the current widely used split or inactivated vaccines to induce cross-clade protective immunity. Immunization with HSP gp96 as adjuvant led to a dramatic increased antigen-specific T cell response to a pandemic H1N1 split vaccine. Notably, gp96 elicited a cross-protective CD8+ T cell response to the internal conserved viral protein NP. Although the split pH1N1 vaccine alone has low cross-protective efficiency, adding gp96 as an adjuvant effectively improved the cross-protection against challenge with a heterologous virus in mice [36]. This reveals the novel property of gp96 in boosting the T cell response against conserved epitopes of influenza virus and its potential use as an adjuvant for human prepandemic inactivated influenza vaccines against different viral subtypes.

### 7.3.2 HSP-Adjuvanted Human Papillomavirus (HPV) Vaccines

HPV infection is strongly associated with cervical cancer, especially for HPV types 16 and 18. Constitutive expression of the viral oncoproteins E6 and E7 is observed in the majority of cervical tumor cells. HSP110, a major HSP of eukaryotic and mammalian cells, could induce HPV E7 epitope-specific T cell response in C57BL/6 and HLA-A2 transgenic mouse model. In addition, HSP110 complexed with E7 epitope elicited stronger ex vivo and in vivo antitumor responses than emulsified complete Freund's adjuvant vaccine [37]. Besides HSP110, calreticulin, HSP70, and gp96 could also act as potent immunoadjuvant to enhance antigen-specific antiviral T cell immunity. The recombinant HPV16 E7-gp96 or calreticulin N terminal fusion proteins induce higher E7-specific and IFN-$\gamma$+T cell response compared to E7 protein alone, greatly delaying the tumor occurrence and growth, and generating potent antiangiogenic effects in mice [38, 39].

### 7.3.3 HSP-Adjuvanted BCG Vaccines

Tuberculosis (TB) continues to pose a serious threat to public health and inflict enormous economic burden to the society despite major progress in therapies. *Mycobacterium tuberculosis* (M.tb), which infected around two billion people worldwide, is a major etiological factor in the development of TB. The BCG vaccine, an intradermal vaccine using live attenuated *Mycobacterium bovis* bacillus, is the most wildly used vaccine covering more than 80% of populations in regions endemic for TB. However, its efficacy in preventing TB varies from 0% to 80%, which may be dependent on patient age and immune status, TB location and the geographic area [40]. As in the natural course of M.tb infection, robust pathogen-specific CD4+, CD8+ T and Th17 cell responses are observed during self-limited or latent infections, whereas active infection is characterized by only weak and impaired T cell responses; T cells are believed to play a critical role in the control of M.tb infection and TB disease [3].

Two major types of adjuvants are available for BCG vaccines, i.e., aluminum salts-based adjuvants and oil-in-water emulsion MF59. Alum adjuvant improves antibody immune responses and the polarized Th2 cell response, and MF59 promotes a mixed Th1 and Th2 cell response. However, their capability to stimulate a cellular immune response is rather limited. There was no evidence of induction of the cellular-mediated immune response and protective potential of using these adjuvants in BCG vaccine [41–43]. Immunization with gp96 adjuvant could induce a significantly increased number of antigen-specific T cells and IFN-$\gamma$-producing CD4+ and CD8+ T cells. The adjuvant effects of gp96 were also reflected by enhanced secretion of the Th1-type cytokines by splenocytes from gp96-immunized mice. The superior T cell immune responses induced with the aid of gp96 correlated with improved protection against challenge with BCG infection (S. Meng, unpublished data). This reveals the novel property of gp96 in boosting T cell responses against mycobacteria infection and its potential use as an adjuvant for BCG vaccine.

## 7.4 HSP-Based Therapeutic Vaccines Against Pathogens

### *7.4.1 Enhancement of gp96-Mediated T Cell Responses*

Despite the immune-regulatory and adjuvant activities of HSP gp96 which induces both innate and adaptive immunity, the effectiveness of gp96-based immunotherapy has been limited. The mechanisms of gp96 activity are still not fully understood, and evidence from different sources has indicated several alternative mechanisms for gp96-induced T cell responses. The internalization of gp96-peptide complexes, which is essential for cross-presentation of antigenic peptides and T cell activation, may be CD91 independent, and that heparin sulfate proteoglycans play an important role in the surface binding of gp96 [44–46]. Moreover, some tumor cells constitutively express receptor associated protein (RAP) that can bind CD91 with high affinity and thereby competitively block its association with gp96 [47]. Given the apparent dependence of gp96-mediated immune responses on the uptake and internalization of gp96 into APCs, it is critical to explore ways to improve these processes and enhance the capacity of gp96 in antigen presentation.

The TAT protein transduction domain (PTD), derived from the HIV-1 transactivator of transcription (TAT) protein, is a short basic region comprising residues 49–57 (RKKRRQRRR), which has been shown to mediate protein transduction in both mice and cultured cells [48]. Many TAT fusion proteins have been generated to deliver a wide variety of size-independent molecules into cells, including peptides, proteins, antisense oligonucleotides, large iron beads, and liposomes [49–52]. The fusion of gp96 with TAT peptide can significantly improve the internalization of gp96 into macrophages and produce dramatic increases of gp96-mediated HBV-specific CTL responses and antiviral efficiency in HBV transgenic mice. Furthermore, the addition of TAT also enhanced gp96-induced antitumor T cell immunity in the B16 melanoma model, supporting the hypothesis that efficient transduction and internalization of gp96-peptide complexes into APCs determine the outcome of gp96-based immunotherapy. Such enhancement of gp96 internalization and its capacity for antigen presentation could promote gp96-mediated CTL responses [53]. This helps to design a more efficient approach to improve the immune activity for this unique T cell adjuvant.

### *7.4.2 Gp96-Based Therapeutic Vaccine Against HBV*

More than 350 million people worldwide are chronically infected with HBV. Chronic HBV infection continues to be a major public health problem as around 15–40% of infected patients will develop life-threatening complications such as cirrhosis, liver failure and hepatocellular carcinoma (HCC). Broad repertoire and strong magnitude of HBV-specific T cell responses are thought to play key roles for virus control and clearance [54]. Immunization with combined HBsAg and HBcAg formulation along with gp96 led to a marked enhancement in antibody and cellular responses toward both HBsAg and HBcAg in HBV transgenic mice. The superior immune responses induced with the aid of gp96 correlated with the improved antiviral effect by vaccination with HBsAg and HBcAg. Immunization with gp96 adjuvant vaccine reduced serum HBsAg level and HBcAg expression in liver tissue by 45% and 90% at maximum, respectively, and decreased serum HBV-DNA level by more than 1000-fold to below or close to the lowest detection limit. Treatment with gp96 elicited an overall 30–40% decrease of Tregs which negatively regulate cellular and antibody immunity during HBV infection [53]. This study reveals the novel property of gp96 in immune modulation and its potential use for breaking immunotolerance in immunotherapy of chronic HBV infection.

### *7.4.3 HSP70-Based Therapeutic Vaccine Against HSV-2*

Herpes simplex virus (HSV)-2, which infects around half a billion people between the ages of 15 and 49, is the leading cause of genital ulcer diseases worldwide. Studies from both mice and humans have demonstrated that viral structural and nonstructural proteins are major targets of HSV-specific CD4+ and CD8+ T cells, serving as candidates to be incorporated in herpes vaccine candidates [55]. Human Hsc70 protein complexed with HSV-2 peptides was tested for safety and immunogenicity in a Phase I clinical trial. The HSP70-based vaccine was well tolerated and safe. All immunized participants demonstrated a statistically significant CD4+ T cell response to HSV-2 antigens, and the vaccine induced a statistically significant CD8+ T cell response as well. This is the first candidate vaccine against HSV-2 which induces a broad CD4+ and CD8+ T cell response in HSV-2 positive participants, and the first HSP-based vaccine to elicit immune responses against viral antigens in humans [56].

## 7.5 HSP as a Target in Chronic Pathogen Infection-Associated Cancer

### *7.5.1 HSP gp96 as a Target in Chronic HBV Infection and Hepatocellular Carcinoma (HCC)*

Numerous preclinical and clinical studies demonstrate the pathogenic roles of chronic HBV infection and inflammation in HCC [57–59]. In chronic hepatitis B, protumorigenic inflammation, which is characterized by liver-infiltrating Th2 cells, Tregs, and M2 macrophages, as well as elevated TNF-α, IL6, IL-1α, and IL-1β expression, may induce persistent hepatocyte regeneration and survival, increasing the neoplastic transformation of hepatocytes [60, 61]. Several inflammation-related signaling pathways are involved in hepatocarcinogenesis, including the NF-κB, JAK-STAT, Raf/MAPK/ERK, Wnt-β-catenin, IRAK-1, and PI3K/AKT/mTOR pathways. Uncovering specific targets of these intertwining pathways may assist in developing more efficient liver cancer treatments.

HSP gp96 mainly functions as a molecular chaperone that participates in the folding and biogenesis of target proteins and guides their assembly and maturation [62, 63]. The normally ER-resident gp96 translocates to the cell membrane under certain circumstances. For example, some microbial stimuli such as *Listeria monocytogenes* and *Escherichia coli* K1 infections upregulate the cell membrane expression of gp96 [64–66]. Aminoacyl tRNA synthetase complex-interacting multifunctional protein 1 (AIMP1) may play a critical role in regulating cell membrane expression of gp96 by affecting the interaction between gp96 and KDELR1 in the ER [67, 68]. Cell membrane gp96 displays different roles in different contexts, such as a bacterial receptor during infections and the incentive for autoimmune diseases [69]. Notably, CD24 was found to be involved in gp96-driven autoimmune disease through regulation of myeloid-derived suppressor cells [70].

HSP gp96 is significantly upregulated in chronic hepatitis B and HCC. Chronic inflammation-mediated regulation of gp96 expression requires a NF-κB cis-regulatory element on the gp96 promoter. Lipopolysaccharide (LPS), muramyl dipeptide (MDP), tumor necrosis factor (TNF), and IL-4 could all induce gp96 expression in macrophages and dendritic cells [71]. HBV × protein also promotes gp96 expression through NF-κB activation, and increased gp96 can in turn promote HBV replication, providing new insights into the regulatory network between gp96 and chronic inflammation/HBV infection [72]. Liver-specific gp96 deletion in mice resulted in a tremendous growth disadvantage in that the remaining small percentage of gp96(+) hepatocytes regenerated disproportionally, rendering them more predisposed to carcinogenesis by diethylnitrosamine (DENA) treatment [73]. In addition, elevated gp96 expression and gp96 translocation to cell membrane are significantly correlated with tumor metastasis and poor prognosis in HBV-infected liver cancer patients.

Mechanistic studies revealed that gp96 promotes p53 degradation through increasing Mdm2 E3 ligase activity, indicating antiapoptotic activity of gp96 [74]. Moreover, gp96 also chaperones multiple strategically important oncogenic clients such as integrin, Wnt coreceptor, IGF1, and TLR [73]. Importantly, cell membrane gp96 directly binds to uPAR and HER2, stabilizes these oncoproteins, and thereby increases their downstream signaling. Targeting cell membrane gp96 with the monoclonal antibody or inhibitor suppress uPAR or HER2-driven cell growth, survival, and invasion [75, 76]. Moreover, the C-terminal domain of cell membrane gp96 directly interacts with estrogen receptor (ER)-α36 on the cell membrane of tumor cells. This interaction stabilizes the ER-α36 protein, thereby increasing its signaling, which, in turn, increases tumor cell growth and invasion. Targeting mgp96 with siRNA or monoclonal antibody blocks the mgp96-ER-α36 interaction and inhibits cancer growth and invasion both in vitro and in vivo [77]. These results validate that cell membrane gp96 is a potential therapeutic target for chronic HBV infection-induced HCC.

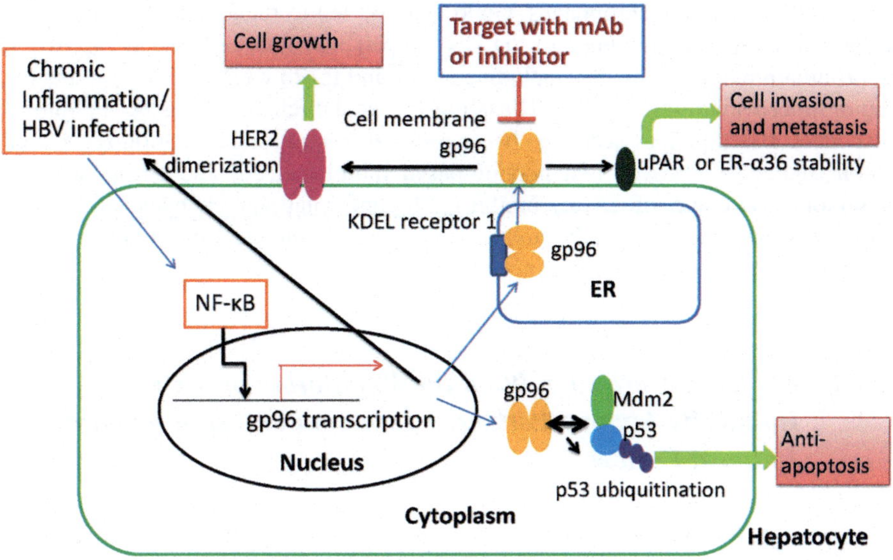


**Fig. 7.1** Chronic HBV infection and inflammation-induced gp96 expression and cell membrane translocation plays an important role in HCC development, growth and metastasis. Elevated gp96 promotes p53 degradation, indicating antiapoptotic activity of gp96. Importantly, cell membrane gp96 directly binds to uPAR and HER2, stabilizes these oncoproteins and thereby increases their downstream signaling. The monoclonal antibody or inhibitor to target cell membrane gp96 greatly suppresses liver tumor growth and metastasis, validating that cell membrane gp96 is a potential therapeutic target for chronic HBV infection and liver cancer

Considering the key role of cell membrane gp96 in the regulation of chronic HBV infection-induced HCC apoptosis, metastasis and growth, gp96 may be an attractive target in chronic HBV infection, which has several implications. First, it may help to address the underlying mechanism of antiapoptotic characteristics in HBV-infected liver cancer. Second, cell membrane gp96 may function as a scaffolding protein to increase uPAR stability and facilitate HER2 dimerization on the cell membrane. Third, more importantly, the cell membrane localization of the normally ER-resident gp96 may serve as a potential negative prognostic marker for chronic hepatitis B and HCC. Given that cell membrane gp96 exists predominantly in malignant but not in normal cells, targeting cell membrane gp96 might provide a novel therapeutic approach for chronic HBV infection and HCC (Fig. 7.1).

## 7.5.2 *HSP gp96 as a Target in Inflammation-Associated Colon Cancers*

Colon carcinogenesis can be driven by infection of microbes. Macrophages play a key role in the development of inflammation-associated colon cancers. Reduced colitis and inflammation-associated colon tumorigenesis were observed in gp96 specific KO mice in macrophages. Gp96 deletion led to reduced mutation rates of β-catenin, increased efficiency of the DNA repair machinery. Reduced expression of proinflammatory cytokines, including IL-17 and IL-23 were also observed in the tumor microenvironment [78]. Thus, the molecular chaperone gp96 in tumor-associated macrophages seems to be involved in driving inflammation-associated colon cancer [79]. Given that peptide-based inhibitors [80] and purine scaffold inhibitors [81] could effectively inhibit gp96 conformational changes and chaperone functions, targeting gp96 may provide a therapeutic opportunity against inflammation-associated colon cancers.

### 7.5.3 Placenta-Derived gp96 in the Regulation of T-Cell Immunity Against Pathogen Infection and Inflammation-Induced Cancer

Chronic infection and inflammation was estimated to contribute to occurrence of 20% of all cancers, such as chronic HBV or HCV infection induced HCC, *Heliobacter pylori*-induced gastric cancer, microbiota infection and inflammation-mediated colon cancer, and HPV (human papillomavirus) infection-induced cervical cancer. Epidemiological studies of somatic mutations show that around 30% of human malignancies are linked to tobacco use, 35% to diet, 14–20% to obesity, 18% to infectious agents, and 7% to radiation or environmental pollutants. Turning the chronic precancerous inflammatory microenvironment (e.g., Th2-polarized immunity) into an anticancer microenvironment (e.g., Th1-polarized immunity) seems to be an attractive approach for cancer prevention and therapy [82].

It is well documented that chronic pathogen infection-mediated development of cancer undergo a long period of incubation time, in the matter of years or even longer, during which the surviving preneoplastic cells have to accumulate mutations to escape the immunosurveillance mechanism, or maintain dormant state via cell cycle arrest to resist elimination [83]. Most neoplastic cells and tumors express embryonic antigens to some extent, which are called carcinoembryonic antigens. Normal cells usually gain embryonic phenotype during neoplastic progression in developing tumors, and the expression or reexpression of embryonic genes is involved in this process. Due to the striking similarity of antigen expression pattern between cancer and embryonic tissues, immunization with embryonic material or cancer stem cells in mice could effectively inhibit transplantable tumor growth and prevent tumorigenesis and carcinogenesis caused by viral and chemical agents [84–87].

Similar to other fetal tissues, the placenta also displays high carcinoembryonic antigens, such as IGF2, HIG-2-a, GPC3, pregnancy-associated plasma protein A (PAPP-A), and MUC1 [88]. Immunization with placental gp96 induced significant and long-term antitumor T-cell immunity in B16-F10 melanoma and TUBO breast cancer mice models. Of note, placental gp96 elicited total protection against 7, 12-dimethylbenz(a)-anthracene (DMBA)-induced mammary tumors in rats, diethylnitrosamine (DEN)-induced HCC and azoxymethane (AOM)/dextran sodium sulfate (DSS)-induced colon cancer. The antitumor activity of placental gp96 was further observed in HER2 transgenic mice. Mechanistic studies revealed that placental gp96 bound to HER2- and MUC1-derived epitopes and activated tumor antigen-specific T-cell responses ([89] and Meng S, unpublished data]. Placenta-derived gp96 could be therefore used as a multivalent cancer vaccine for both preventive and therapeutic purposes. Deep and comprehensive analyses of the peptide repertoire which is associated with placenta-derived gp96 are needed to further dissect its T-cell-mediated immune responses against various tumors and its use as a potential prophylactic cancer vaccine in humans, especially those with a high risk for developing chronic infection-associated neoplastic diseases.

## 7.6 Conclusion and Perspective

Since its discovery in the chemically induced sarcomas as a tumor rejection antigen more than 20 years ago [90], the autologus gp96, purified from cancers or pathogen-infected cells, as prophylactic and therapeutic vaccines has been extensively studied in both animal models and human clinical trials. However, the effectiveness of gp96-based vaccines in immunotherapy so far seems to be limited probably due to immune tolerance, including Tregs and the inhibitory immune receptors such as CTLA4 and PD-1/PD-L1, as well as lack of molecularly defined appropriate antigens for better vaccine optimization.

There is now compelling evidence that gp96 is involved in diverse aspects of innate and adaptive immunomodulation. As immune checkpoint inhibitors against CTLA-4, PD-1 pathway and others have emerged as a successful treatment approach for patients with advanced cancer, a combination strategy of HSP vaccination with means to breaking immune tolerance may prove to be a fruitful future direction for both cancers and chronic infectious diseases. In addition, it is also worthwhile to further explore the mechanisms and feasibility of placental gp96 as multivalent prophylactic and therapeutic cancer vaccine, as well as high-dose gp96-induced Tregs in treatment of pathogen-associated immune hyperactivation and autoimmune diseases.

Finally, the intrinsic roles of HSPs including gp96 in oncogenesis and viral infections have been elevated to the center stage. Detailed molecular and mechanistic studies of gp96 and its client network in these settings will undoubtedly create more opportunities for developing gp96-targeted therapies against cancer and infectious diseases.

## References


1. Tan AT, Loggi E, Boni C, Chia A, Gehring AJ, Sastry KS et al (2008) Host ethnicity and virus genotype shape the hepatitis B virus-specific T-cell repertoire. J Virol 82(22):10986–10997
2. Tan AC, Deliyannis G, Bharadwaj M, Brown LE, Zeng W, Jackson DC (2013) The design and proof of concept for a CD8(+) T cell-based vaccine inducing cross-subtype protection against influenza A virus. Immunol Cell Biol 91(1):96–104
3. Jasenosky LD, Scriba TJ, Hanekom WA, Goldfeld AE (2015) T cells and adaptive immunity to Mycobacterium tuberculosis in humans. Immunol Rev 264(1):74–87
4. Randazzo M, Terness P, Opelz G, Kleist C (2012) Active-specific immunotherapy of human cancers with the heat shock protein Gp96-revisited. Int J Cancer 130(10):2219–2231
5. Crane CA, Han SJ, Ahn B, Oehlke J, Kivett V, Fedoroff A et al (2013) Individual patient-specific immunity against high-grade glioma after vaccination with autologous tumor derived peptides bound to the 96 KD chaperone protein. Clin Cancer Res 19(1):205–214
6. Strbo N, Vaccari M, Pahwa S, Kolber MA, Doster MN, Fisher E et al (2013) Cutting edge: novel vaccination modality provides significant protection against mucosal infection by highly pathogenic simian immunodeficiency virus. J Immunol 190(6):2495–2499
7. Kropp LE, Garg M, Binder RJ (2010) Ovalbumin-derived precursor peptides are transferred sequentially from gp96 and calreticulin to MHC class I in the endoplasmic reticulum. J Immunol 184(10):5619–5627
8. Robert J, Ramanayake T, Maniero GD, Morales H, Chida AS (2008) Phylogenetic conservation of glycoprotein 96 ability to interact with CD91 and facilitate antigen cross-presentation. J Immunol 180(5):3176–3182
9. Matsutake T, Sawamura T, Srivastava PK (2010) High efficiency CD91- and LOX-1-mediated re-presentation of gp96-chaperoned peptides by MHC II molecules. Cancer Immun 10:7
10. Messmer MN, Pasmowitz J, Kropp LE, Watkins SC, Binder RJ (2013) Identification of the cellular sentinels for native immunogenic heat shock proteins in vivo. J Immunol 191(8):4456–4465
11. Yang Y, Liu B, Dai J, Srivastava PK, Zammit DJ, Lefrancois L et al (2007) Heat shock protein gp96 is a master chaperone for toll-like receptors and is important in the innate function of macrophages. Immunity 26(2):215–226
12. Wu S, Hong F, Gewirth D, Guo B, Liu B, Li Z (2012) The molecular chaperone gp96/GRP94 interacts with Toll-like receptors and integrins via its C-terminal hydrophobic domain. J Biol Chem 287(9):6735–6742
13. Pawaria S, Binder RJ (2011) CD91-dependent programming of T-helper cell responses following heat shock protein immunization. Nat Commun 2:521
14. Hornung V, Bauernfeind F, Halle A, Samstad EO, Kono H, Rock KL et al (2008) Silica crystals and aluminum salts activate the NALP3 inflammasome through phagosomal destabilization. Nat Immunol 9(8):847–856
15. Lambrecht BN, Kool M, Willart MA, Hammad H (2009) Mechanism of action of clinically approved adjuvants. Curr Opin Immunol 21(1):23–29
16. Coffman RL, Sher A, Seder RA (2010) Vaccine adjuvants: putting innate immunity to work. Immunity 33(4):492–503
17. Campbell DJ (2015) Control of regulatory T cell migration, function, and homeostasis. J Immunol 195(6):2507–2513
18. Li H, Zhou M, Han J, Zhu X, Dong T, Gao GF et al (2005) Generation of murine CTL by a hepatitis B virus-specific peptide and evaluation of the adjuvant effect of heat shock protein glycoprotein 96 and its terminal fragments. J Immunol 174(1):195–204
19. Liu Z, Li X, Qiu L, Zhang X, Chen L, Cao S et al (2009) Treg suppress CTL responses upon immunization with HSP gp96. Eur J Immunol 39(11):3110–3120
20. Zhang Y, Wu BX, Metelli A, Thaxton JE, Hong F, Rachidi S et al (2015) GP96 is a GARP chaperone and controls regulatory T cell functions. J Clin Invest 125(2):859–869

21. Zhang Y, Ansa-Addo E, Li Z (2015) GP96: safeguarding Treg. Oncotarget 6(24):19936–19937
22. Chen Q, Davidson TS, Huter EN, Shevach EM (2009) Engagement of TLR2 does not reverse the suppressor function of mouse regulatory T cells, but promotes their survival. J Immunol 183(7):4458–4466
23. Li X, Liu Z, Yan X, Zhang X, Li Y, Zhao B et al (2013) Induction of regulatory T cells by high-dose gp96 suppresses murine liver immune hyperactivation. PLoS One 8(7):e68997
24. Li J, Qiu SJ, She WM, Wang FP, Gao H, Li L et al (2012) Significance of the balance between regulatory T (Treg) and T helper 17 (Th17) cells during hepatitis B virus related liver fibrosis. PLoS One 7(6):e39307
25. Buckwalter MR, Srivastava PK (2008) "It is the antigen(s), stupid" and other lessons from over a decade of vaccitherapy of human cancer. Semin Immunol 20(5):296–300
26. Yan X, Zhang X, Wang Y, Li X, Wang S, Zhao B et al (2011) Regulatory T-cell depletion synergizes with gp96-mediated cellular responses and antitumor activity. Cancer Immunol Immunother 60(12):1763–1774
27. Rech AJ, Vonderheide RH (2009) Clinical use of anti-CD25 antibody daclizumab to enhance immune responses to tumor antigen vaccination by targeting regulatory T cells. Ann N Y Acad Sci 1174:99–106
28. Jacobs JF, Punt CJ, Lesterhuis WJ, Sutmuller RP, Brouwer HM, Scharenborg NM et al (2010) Dendritic cell vaccination in combination with anti-CD25 monoclonal antibody treatment: a phase I/II study in metastatic melanoma patients. Clin Cancer Res 16(20):5067–5078
29. Lee YN, Lee YT, Kim MC, Gewirtz AT, Kang SM (2016) A novel vaccination strategy mediating the induction of lung-resident memory CD8 T cells confers heterosubtypic immunity against future pandemic influenza virus. J Immunol 196(6):2637–2645
30. Uddback IE, Pedersen LM, Pedersen SR, Steffensen MA, Holst PJ, Thomsen AR et al (2016) Combined local and systemic immunization is essential for durable T-cell mediated heterosubtypic immunity against influenza A virus. Sci Rep 6:20137
31. Zhang J, Fan HY, Zhang Z, Huang JN, Ye Y, Liao M (2016) Recombinant baculovirus vaccine containing multiple M2e and adjuvant LTB induces T cell dependent, cross-clade protection against H5N1 influenza virus in mice. Vaccine 34(5):622–629
32. Stanekova Z, Vareckova E (2010) Conserved epitopes of influenza A virus inducing protective immunity and their prospects for universal vaccine development. Virol J 7:351
33. Huang B, Wang W, Li R, Wang X, Jiang T, Qi X et al (2012) Influenza A virus nucleoprotein derived from Escherichia coli or recombinant vaccinia (Tiantan) virus elicits robust cross-protection in mice. Virol J 9:322
34. Ichinohe T, Ainai A, Tashiro M, Sata T, Hasegawa H (2009) PolyI:polyC12U adjuvant-combined intranasal vaccine protects mice against highly pathogenic H5N1 influenza virus variants. Vaccine 27(45):6276–6279
35. Wang X, Zhang W, Liu F, Zheng M, Zheng D, Zhang T et al (2012) Intranasal immunization with live attenuated influenza vaccine plus chitosan as an adjuvant protects mice against homologous and heterologous virus challenge. Arch Virol 157(8):1451–1461
36. Ju Y, Fan H, Liu J, Hu J, Li X, Li C et al (2014) Heat shock protein gp96 adjuvant induces T cell responses and cross-protection to a split influenza vaccine. Vaccine 32(23):2703–2711
37. Ding Z, Ou R, Ni B, Tang J, Xu Y (2013) Cytolytic activity of the human papillomavirus type 16 E711-20 epitope-specific cytotoxic T lymphocyte is enhanced by heat shock protein 110 in HLA-A*0201 transgenic mice. Clin Vaccine Immunol 20(7):1027–1033
38. Mohit E, Bolhassani A, Zahedifard F, Taslimi Y, Rafati S (2012) The contribution of NT-gp96 as an adjuvant for increasing HPV16 E7-specific immunity in C57BL /6 mouse model. Scand J Immunol 75(1):27–37
39. Liu B, Ye D, Song X, Zhao X, Yi L, Song J et al (2008) A novel therapeutic fusion protein vaccine by two different families of heat shock proteins linked with HPV16 E7 generates potent antitumor immunity and antiangiogenesis. Vaccine 26(10):1387–1396
40. Principi N, Esposito S (2015) The present and future of tuberculosis vaccinations. Tuberculosis (Edinb) 95(1):6–13

41. Agger EM (2016) Novel adjuvant formulations for delivery of anti-tuberculosis vaccine candidates. Adv Drug Deliv Rev 102:73–82
42. O'Hagan DT, Ott GS, De Gregorio E, Seubert A (2012) The mechanism of action of MF59 – an innately attractive adjuvant formulation. Vaccine 30(29):4341–4348
43. Graves AJ, Hokey DA (2015) Tuberculosis vaccine development: Shifting focus amid increasing development challenges. Hum Vaccin Immunother 11(8):1910–1916
44. Berwin B, Hart JP, Pizzo SV, Nicchitta CV (2002) Cutting edge: CD91-independent cross-presentation of GRP94(gp96)-associated peptides. J Immunol 168(9):4282–4286
45. Jockheck-Clark AR, Bowers EV, Totonchy MB, Neubauer J, Pizzo SV, Nicchitta CV (2010) Re-examination of CD91 function in GRP94 (glycoprotein 96) surface binding, uptake, and peptide cross-presentation. J Immunol 185(11):6819–6830
46. Lev A, Dimberu P, Das SR, Maynard JC, Nicchitta CV, Bennink JR et al (2009) Efficient cross-priming of antiviral CD8+ T cells by antigen donor cells is GRP94 independent. J Immunol 183(7):4205–4210
47. Pawaria S, Messmer MN, Zhou YJ, Binder RJ (2011) A role for the heat shock protein-CD91 axis in the initiation of immune responses to tumors. Immunol Res 50(2-3):255–260
48. Schwarze SR, Dowdy SF (2000) In vivo protein transduction: intracellular delivery of biologically active proteins, compounds and DNA. Trends Pharmacol Sci 21(2):45–48
49. Wang S, Han Q, Zhang N, Chen J, Liu Z, Zhang G et al (2010) HBcAg18-27 epitope fused to HIV-Tat 49-57 adjuvanted with CpG ODN induces immunotherapeutic effects in transgenic mice. Immunol Lett 127(2):143–149
50. Chu X, Wu B, Fan H, Hou J, Hao J, Hu J et al (2016) PTD-fused p53 as a potential antiviral agent directly suppresses HBV transcription and expression. Antivir Res 127:41–49
51. Nam HY, Kim J, Kim S, Yockman JW, Kim SW, Bull DA (2011) Cell penetrating peptide conjugated bioreducible polymer for siRNA delivery. Biomaterials 32(22):5213–5222
52. Krpetic Z, Saleemi S, Prior IA, See V, Qureshi R, Brust M (2011) Negotiation of intracellular membrane barriers by TAT-modified gold nanoparticles. ACS Nano 5(6):5195–5201
53. Zhao B, Wang Y, Zhang Y, Li Y, Zhang X, Xu Y et al (2013) TAT-mediated gp96 transduction to APCs enhances gp96-induced antiviral and antitumor T cell responses. Vaccine 31(3):545–552
54. Knolle PA, Thimme R (2014) Hepatic immune regulation and its involvement in viral hepatitis infection. Gastroenterology 146(5):1193–1207
55. Kuo T, Wang C, Badakhshan T, Chilukuri S, BenMohamed L (2014) The challenges and opportunities for the development of a T-cell epitope-based herpes simplex vaccine. Vaccine 32(50):6733–6745
56. Wald A, Koelle DM, Fife K, Warren T, Leclair K, Chicz RM et al (2011) Safety and immunogenicity of long HSV-2 peptides complexed with rhHsc70 in HSV-2 seropositive persons. Vaccine 29(47):8520–8529
57. Potikha T, Stoyanov E, Pappo O, Frolov A, Mizrahi L, Olam D et al (2013) Interstrain differences in chronic hepatitis and tumor development in a murine model of inflammation-mediated hepatocarcinogenesis. Hepatology 58(1):192–204
58. Nakagawa H, Maeda S (2012) Inflammation- and stress-related signaling pathways in hepatocarcinogenesis. World J Gastroenterol 18(31):4071–4081
59. Bishayee A (2014) The role of inflammation and liver cancer. Adv Exp Med Biol 816:401–435
60. Su B, Luo T, Zhu J, Fu J, Zhao X, Chen L et al (2015) Interleukin-1beta/Iinterleukin-1 receptor-associated kinase 1 inflammatory signaling contributes to persistent Gankyrin activation during hepatocarcinogenesis. Hepatology 61(2):585–597
61. DiDonato JA, Mercurio F, Karin M (2012) NF-kappaB and the link between inflammation and cancer. Immunol Rev 246(1):379–400
62. Marzec M, Eletto D, Argon Y (2012) GRP94: an HSP90-like protein specialized for protein folding and quality control in the endoplasmic reticulum. Biochim Biophys Acta 1823(3):774–787
63. Hong F, Liu B, Chiosis G, Gewirth DT, Li Z (2013) Alpha7 helix region of alphaI domain is crucial for integrin binding to endoplasmic reticulum chaperone gp96: a potential therapeutic target for cancer metastasis. J Biol Chem 288(25):18243–18248

64. Martins M, Custodio R, Camejo A, Almeida MT, Cabanes D, Sousa S (2012) Listeria monocytogenes triggers the cell surface expression of Gp96 protein and interacts with its N terminus to support cellular infection. J Biol Chem 287(51):43083–43093
65. Mittal R, Prasadarao NV (2011) gp96 expression in neutrophils is critical for the onset of Escherichia coli K1 (RS218) meningitis. Nat Commun 2:552
66. Cabanes D, Sousa S, Cebria A, Lecuit M, Garcia-del Portillo F, Cossart P (2005) Gp96 is a receptor for a novel Listeria monocytogenes virulence factor, Vip, a surface protein. EMBO J 24(15):2827–2838
67. Han JM, Park SG, Liu B, Park BJ, Kim JY, Jin CH et al (2007) Aminoacyl-tRNA synthetase-interacting multifunctional protein 1/p43 controls endoplasmic reticulum retention of heat shock protein gp96: its pathological implications in lupus-like autoimmune diseases. Am J Pathol 170(6):2042–2054
68. Kim G, Han JM, Kim S (2010) Toll-like receptor 4-mediated c-Jun N-terminal kinase activation induces gp96 cell surface expression via AIMP1 phosphorylation. Biochem Biophys Res Commun 397(1):100–105
69. Dai J, Liu B, Ngoi SM, Sun S, Vella AT, Li Z (2007) TLR4 hyperresponsiveness via cell surface expression of heat shock protein gp96 potentiates suppressive function of regulatory T cells. J Immunol 178(5):3219–3225
70. Thaxton JE, Liu B, Zheng P, Liu Y, Li Z (2014) Deletion of CD24 impairs development of heat shock protein gp96-driven autoimmune disease through expansion of myeloid-derived suppressor cells. J Immunol 192(12):5679–5686
71. Wolfram L, Fischbeck A, Frey-Wagner I, Wojtal KA, Lang S, Fried M et al (2013) Regulation of the expression of chaperone gp96 in macrophages and dendritic cells. PLoS One 8(10):e76350
72. Fan H, Yan X, Zhang Y, Zhang X, Gao Y, Xu Y et al (2013) Increased expression of Gp96 by HBx-induced NF-kappaB activation feedback enhances hepatitis B virus production. PLoS One 8(6):e65588
73. Rachidi S, Sun S, Wu BX, Jones E, Drake RR, Ogretmen B et al (2015) Endoplasmic reticulum heat shock protein gp96 maintains liver homeostasis and promotes hepatocellular carcinogenesis. J Hepatol 62(4):879–888
74. Wu B, Chu X, Feng C, Hou J, Fan H, Liu N et al (2015) Heat shock protein gp96 decreases p53 stability by regulating Mdm2 E3 ligase activity in liver cancer. Cancer Lett 359(2):325–334
75. Hou J, Li X, Li C, Sun L, Zhao Y, Zhao J et al (2015) Plasma membrane gp96 enhances invasion and metastatic potential of liver cancer via regulation of uPAR. Mol Oncol 9(7):1312–1323
76. Li X, Sun L, Hou J, Gui M, Ying J, Zhao H et al (2015) Cell membrane gp96 facilitates HER2 dimerization and serves as a novel target in breast cancer. Int J Cancer 137(3):512–524
77. Hou J, Deng M, Li X, Liu W, Chu X, Wang J et al (2015) Chaperone gp96 mediates ER-alpha36 cell membrane expression. Oncotarget 6(31):31857–31867
78. Morales C, Rachidi S, Hong F, Sun S, Ouyang X, Wallace C et al (2014) Immune chaperone gp96 drives the contributions of macrophages to inflammatory colon tumorigenesis. Cancer Res 74(2):446–459
79. Hong F, Wu BX, Li Z (2014) Molecular regulation of macrophages in unleashing cancer-related inflammation. Oncoimmunology 3(1):e27659
80. Kliger Y, Levy O, Oren A, Ashkenazy H, Tiran Z, Novik A et al (2009) Peptides modulating conformational changes in secreted chaperones: from in silico design to preclinical proof of concept. Proc Natl Acad Sci U S A 106(33):13797–13801
81. Patel PD, Yan P, Seidler PM, Patel HJ, Sun W, Yang C et al (2013) Paralog-selective Hsp90 inhibitors define tumor-specific regulation of HER2. Nat Chem Biol 9(11):677–684
82. Coussens LM, Zitvogel L, Palucka AK (2013) Neutralizing tumor-promoting chronic inflammation: a magic bullet? Science 339(6117):286–291
83. Uhr JW, Pantel K (2011) Controversies in clinical cancer dormancy. Proc Natl Acad Sci U S A 108(30):12396–12400
84. Brewer BG, Mitchell RA, Harandi A, Eaton JW (2009) Embryonic vaccines against cancer: an early history. Exp Mol Pathol 86(3):192–197

85. Dong W, Du J, Shen H, Gao D, Li Z, Wang G et al (2010) Administration of embryonic stem cells generates effective antitumor immunity in mice with minor and heavy tumor load. Cancer Immunol Immunother 59(11):1697–1705
86. Yaddanapudi K, Mitchell RA, Putty K, Willer S, Sharma RK, Yan J et al (2012) Vaccination with embryonic stem cells protects against lung cancer: is a broad-spectrum prophylactic vaccine against cancer possible? PLoS One 7(7):e42289
87. Duarte S, Momier D, Baque P, Casanova V, Loubat A, Samson M et al (2013) Preventive cancer stem cell-based vaccination reduces liver metastasis development in a rat colon carcinoma syngeneic model. Stem Cells 31(3):423–432
88. Sood R, Zehnder JL, Druzin ML, Brown PO (2006) Gene expression patterns in human placenta. Proc Natl Acad Sci U S A 103(14):5478–5483
89. Zhao B, Wang Y, Wu B, Liu S, Wu E, Fan H et al (2013) Placenta-derived gp96 as a multivalent prophylactic cancer vaccine. Sci Rep 3:1947
90. Srivastava PK, DeLeo AB, Old LJ (1986) Tumor rejection antigens of chemically induced sarcomas of inbred mice. Proc Natl Acad Sci U S A 83(10):3407–3411

# Chapter 8
# An Ancestral Immune Surveillance System in the Amphibian *Xenopus* Connecting Certain Heat Shock Proteins with Classical and Nonclassical MHC Class I Molecules

**Jacques Robert, Maureen Banach, and Eva-Stina Edholm**

**Abstract** Studies in the amphibian *Xenopus*, a vertebrate species that diverged from a common ancestor with mouse and human more than 350 million years ago, provide evolutionary insights into the convergent roles of certain hsps such as gp96 and HSP70 as well as classical and nonclassical MHC class I molecules in cancer immune surveillance. Evidence that in *Xenopus* gp96 and HSP70 can elicit potent antitumor responses dependent on antigen representation by nonclassical MHC class Ib molecules and presumably involving innate T cells suggests the existence of an ancestral immune surveillance system in antigen-presenting cells such as macrophages integrating hsps with classical and nonclassical MHC molecules. The particular connection revealed in *Xenopus* between hsps and nonclassical MHC molecules presenting conserved patterns to innate T cells affords new avenues to develop therapeutic strategies against cancer.

**Keywords** Comparative immunology · Innate T cells · Tumor immunity · Evolution · Unconventional T cells

## 8.1 Introduction

Heat shock proteins (hsps) are evolutionarily ancient and highly conserved molecular chaperones constituting several multigenic families that are produced by all cell types and perform essential biological functions under normal as well as stressful physiological conditions [1]. Some of these hsps including gp96 (a member of the hsp90 family) and the cytosolic 70 kDa hsps or HSP70 (defining indistinctively

J. Robert (✉) · M. Banach · E.-S. Edholm
Department of Microbiology and Immunology, University of Rochester Medical Center, Rochester, NY, USA
e-mail: Jacques_Robert@urmc.rochester.edu

 
R. J. Binder, P. K. Srivastava (eds.), *Heat Shock Proteins in the Immune System*,
https://doi.org/10.1007/978-3-319-69042-1_8

both the inducible hsp72 and the constitutively expressed hsc73) have received a lot of attention because of their potential use in tumor immunotherapy (reviewed in [2–4]). HSP70 and gp96 have been shown to elicit potent CD8 T-cell responses specific against the antigenic peptides they chaperone not only in humans and mice [5–7] but also in frogs [8, 9]. These hsp-mediated CD8 T-cell responses are MHC class I restricted and depend on the internalization of the hsp-antigen complexes by endocytic receptors such as the α2-macroglobulin receptor CD91 at the surface of antigen-presenting cells (APCs; [10, 11]). This is followed by the representation of chaperoned antigenic peptides by MHC class Ia molecules on APCs to CD8 T cells [7, 12, 13]. The functional connection between hsp chaperoning and MHC class I antigen presentation may have even further ramifications than previously thought considering that in addition to classical MHC class Ia (class Ia) a growing number of nonclassical MHC class Ib (class Ib) and class I-like gene have been characterized (reviewed in [14, 15]). Some of these class Ib genes encode proteins that are hypothesized to be indicators of intracellular stress and malignancy (reviewed in [16, 17]. The potential role of these class Ib molecules is of particular relevance in immune surveillance and recognition of aggressive class Ia low or negative tumor cells through their interaction with T-cell receptors and/or non-T-cell inhibitory or triggering receptors expressed by NK and unconventional T cells.

Focusing on two of the most conserved hsps, gp96 and hsp70, studies in the amphibian *Xenopus* have provided compelling evidence that the immunological properties of these molecular chaperones, especially their significant antitumor responses, have been conserved during evolution (Reviewed in [18]. Comparably, while nonclassical MHC class Ib genes in *Xenopus* do not share a direct common ancestor with their mammalian counterparts, some of these genes encode molecules with striking analogous functions including class Ib-restricted unconventional T-cell-mediated antitumor immune responses.

We review here recent advances using the amphibian *Xenopus* to explore the potential of an ancestral immune surveillance system composed of hsps such as gp96 and hsp70, endocytic receptors such as CD91 and classical and nonclassical MHC class I molecules.

## 8.2 The *Xenopus* Immune System

The immune system of the South African clawed frog *Xenopus laevis* exhibits all the basic elements of jawed vertebrate immunity. The primary immune organs thymus and spleen and adaptive B- and T-cell effectors expressing a wide Ig and TCR repertoire generated by RAG-mediated somatic diversification as well as innate cell effectors such as neutrophils and macrophages are all conserved in *Xenopus* (reviewed in [19]). In fact, the fully sequenced and annotated genomes of two different *Xenopus* species, *X. tropicalis* and *X. laevis*, have provided compelling evidence of the remarkably high degree of overall conservation of immune genes between *Xenopus* and human.

One intriguing aspect of anuran amphibians such as *Xenopus* that is not encountered in mammals is that the development of the immune system occurs at two distinct times: first during larval life and then again during the metamorphic transition from tadpole to adult [20, 21]. Specifically, the *Xenopus* thymus is first colonized by embryonic stem cells a few days after fertilization [22]. During metamorphosis, the thymus loses about 90% of its lymphocytes [23]. This loss is followed by a second wave of stem cell immigration [24, 25]. The tadpole is free-swimming and amenable to a variety of surgical (e.g., thymectomy, transplantation) and nonsurgical (e.g., adoptive transfer of leucocytes, injection of hormones, antibodies) interventions. Therefore, studies in *Xenopus* tadpoles can be helpful in collecting valuable information otherwise difficult to gather from in utero studies in mammals (e.g., development of self-tolerance to adult-specific antigens, acquisition of a second T-cell repertoire, and ontogeny of T-cell subsets in a natural setting).

A second aspect of *Xenopus* immunology that makes it attractive as a model is the absence of classical MHC class Ia protein expression in tadpoles until the onset of metamorphosis. Surface class Ia expression is first detected on erythrocytes and on splenic leukocyte populations at pro-metamorphic stages [21, 26, 27]. Although tadpoles are immunocompetent and have CD8 T cells, the larval thymus lacks significant expression of class Ia and LMP7 genes until metamorphosis, which suggests an inefficient class Ia-restricted T-cell education during larval life [21, 28]. Conversely, multiple class Ib genes are expressed by thymocytes at the onset of thymic organogenesis consistent with a role of class Ib molecules in early T-cell development.

Thus, the high degree of functional conservation of the *Xenopus* immune system with human, the natural class Ia-deficient tadpole stages, as well as the amenability of *Xenopus* to in vivo experimentation make it a highly relevant nonmammalian model (reviewed in [19, 29]). In particular, *Xenopus* is well suited to study tumor immune surveillance and as such has proven instrumental to exploring innovative approaches for cancer immunotherapy (reviewed in [19, 30]).

## 8.3 Lymphoid Tumors and Tumor Immunity in *Xenopus*

*X. laevis* is the only amphibian species in which a series of true lymphoid tumor cell lines have been derived and characterized from spontaneously occurring thymic tumors ([31, 32]. Two similar thymic tumors were also reported at the *Xenopus* colony at Tulane University around the same time [33]. More recently, another type of spontaneous leukocytic, possibly monocytic, tumor very different from the thymic tumors originally characterized was described [34].

Importantly, the occurrence of spontaneous thymic tumors in MHC-defined inbred and *X. laevis/X. gilli* isogenetic clones has provided a unique opportunity to derive lymphoid tumor lines growing in in vitro culture as well as in vivo following transplantation in compatible *X. laevis* host [32, 35]. From the partially inbred F strain homozygous of the *f* MHC haplotype, two different tumor lines (B3B7 and ff-2) were derived, whereas from the isogenetic clone LG-15 heterozygous for the MHC haplotype a/c, 15/0 and 15/40 lines were obtained. These cell lines are all nonadherent and grow continuously at 27 °C with a generation time of 18–24 h [36]. All four tumor lines share a mixed immature T-/B-cell phenotype: they all express several pan T-cell markers such as CD8 and CD5 but have also rearranged their Ig gene loci. All the tumor cell lines also express the cortical thymocyte-specific Xenopus cell surface marker (CTX), a marker of immature thymocytes that in the organism is only expressed by cortical thymocytes [37, 38]. Another salient feature exhibited by all these tumor lines is the expression of high level expression of several *Xenopus* nonclassical MHC class Ib (*XNC*) genes, including XNC1, 4, 10, and 11 as well as β2-microglobulin [39]. In contrast, only the ff-2 tumor expresses low levels of classical MHC class Ia at the cell surface, whereas 15/0, 15/40, and B3B7 cell lines are all class Ia-negative [32, 35].

Two of these lymphoid tumor cell lines have remained transplantable in compatible hosts. The ff-2 tumor is transplantable in the MHC homozygous *f/f* partially inbred F strain, whereas the 15/0 can grow in the isogenetic LG-15 background. Interestingly, the ff-2 tumor line is tumorigenic when transplanted into F tadpoles but not into F adults. The rejection of ff-2 tumor in F adults is abrogated by γ-irradiation that preferentially depletes thymocytes and is impaired in T-cell-deficient thymectomized animals, which suggests the critical involvement of adult T cells that differentiate just after metamorphosis [35, 40]. Comparably, the 15/0 tumor cells are highly tumorigenic when transplanted into both tadpole and adult LG-15 hosts [32, 35]. In addition, the 15/0 tumor line is transplantable and tumorigenic in another isogenetic clone, LG-6, that shares the same MHC haplotypes (a/c) with LG-15 animals but differs at multiple minor histocompatibility (H) loci [41]. This difference in minor H-antigens has been instrumental in exploring antigen-specific antitumor immunity in *Xenopus* as delineated in the next chapter.

Initial in vivo and in vitro studies have revealed that in *X. laevis* as in mammals NK and CD8 T cells are critical antitumor cell effectors [41]. Briefly, the involvement of NK cells was demonstrated by anti-NK antibody treatment followed by tumor transplantation assays and by an in vitro cytotoxic assay [41–43]. Thymectomy at early developmental stage before cell precursor immigration and sublethal γ-irradiation that mainly affect dividing thymocytes and circulating T cell provided evidence of CD8 T cells requirement to control malignancy [35, 40, 44]. Importantly, taking advantage of the absence of class Ia expression by 15/0 tumor cells has allowed us to shed light on the unappreciated roles of nonclassical MHC class Ib molecules and unconventional class Ib-restricted T cell in *X. laevis* tumor immunity (see Chap. 5).

## 8.4 Conservation of Antitumor Properties of Heat Shock Proteins

The *X. laevis* tumor immunity model has provided evolutionary evidence of the ability of certain hsps such as the endoplasmic resident gp96 and the cytosolic HSP70 to elicit potent antitumor protective T-cell responses. In mammals, these molecules can induce pro-inflammatory cytokines, stimulate NK cells, and elicit potent cytotoxic CD8 T-cell responses against the antigenic peptides they chaperone [2–4]. The representation of antigens chaperoned by these hsps in the context of MHC class Ia by APCs critically involves the endocytic receptor CD91 [10, 11] as well as other scavenger receptors [45–47]. The additional interaction of these hsps with various signalling receptors such as TLRs is associated with their ability to stimulate inflammation [48, 49].

Given the high degree of evolutionarily conservation of gp96 and hsp70 across vertebrate and even invertebrate species, it was of interest to determine whether the immunostimulatory properties of these hsps, especially regarding antitumor immunity, were also conserved in amphibians such as *Xenopus*. Using minor H-Ags differences between LG-15 and LG-6 cloned frogs, it was first demonstrated that, as in mouse and human, both gp96 and hsp70 were able to represent chaperoned minor H-Ags and generate efficient CD8 T-cell responses recognizing and killing targets expressing the same minor H-Ags in a MHC-restricted fashion [8]. Immunization by direct subcutaneous injection of hsp70 or gp96 chaperoning minor H-Ags as well as by adoptive transfer of macrophages pulsed with hsp70/gp96-minor H-Ag complexes was shown to generate immunological memory to minor H-Ags leading to accelerated rejection of minor H-Ag-matched skin grafts [8, 50]. As in mammals, *Xenopus* gp96 and HSP70 can interact with the endocytic receptor CD91 at the surface of APCs, which leads to its rapid internalization and the representation of its bound antigens by MHC class Ia [51]. These studies in *Xenopus* strongly suggest that certain hsps (gp96, HSP70) and hsp receptors (CD91) are all integral parts of an ancestral system of immune surveillance. The importance of this system in controlling neoplasia is highlighted by its conservation for more than the 350 million years that separate amphibian and mammals from their common ancestor.

Furthermore, since, in contrast to skin grafts, the 15/0 lymphoid tumor does not express class Ia molecules, our comparative tumor immunity model has permitted investigation of the potential roles of hsps in stimulating MHC class Ia-unrestricted NK and unconventional T cells in the context of antitumor immunity. Both in vivo and in vitro studies demonstrated that immune responses against 15/0 tumor cells in *X. laevis* involve NK cells and unconventional classical class Ia-unrestricted CD8 cytotoxic T cells (CCU-CTLs) that both were shown to kill 15/0 tumor cells but not class Ia expressing non-tumoral lymphoblast targets in vitro [41]. The critical involvement of chaperoned antigens in hsp-mediated anti-15/0 tumor immune responses in the absence of class Ia presentation is supported by several lines of evidence. For both gp96 and hsp70, native forms purified from non-tumoral organs (e.g., liver) or recombinant forms

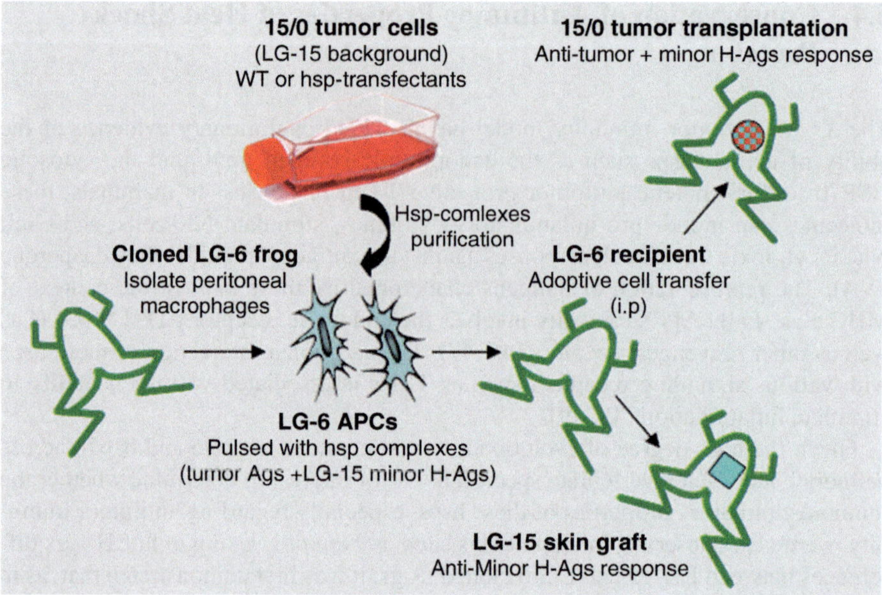


**Fig. 8.1** Schematic of the antigen representation assay developed in *Xenopus*. Peritoneal macrophages elicited by stimulation with heat-killed *E. coli* are recovered from LG-6 adults by peritoneal lavage and used as APCs. Hsps are purified from 15/0 tumor WT or stable transfectant expressing tagged recombinant *Xenopus* hsps. Since 15/0 tumor is on the LG-15 background, hsps chaperone both minor H and tumor Ags. LG-6 macrophages are pulsed for 1 h on ice with the hsp complexes at a concentration of 0.5–1 mg per $1 \times 10^5$ cells, extensively washed, and then adoptively transferred into LG-6 recipients ($5 \times 10^5$ cells per animal). Hsp-mediated immune responses elicited against minor H-Ags can be monitored in vivo by monitoring the rejection time of minor H-disparate LG-15 skin graft. Hsp-mediated antitumor immune response can be monitored by determining the time of tumor appearance following injection of 15/0 tumors

produced from bacteria or non-15/0 cells (e.g., B3B7 cells) did not elicit significant anti-15/0 tumor immune response and the removal of ligands from hsp70 by ADP abrogated anti-15/0 immunogenicity [9, 50].

To specifically address MHC class Ia-dependent and class Ia-independent antigen representation, we developed an in vivo adoptive cell transfer assay using *X. laevis* peritoneal macrophage (pMac) as APCs that is depicted in Fig. 8.1. First, we demonstrated that adoptive transfer of pMac exposed to either gp96- or hsp70-minor H-Ags complexes generated a CD8 T-cell response specifically against minor H-skin Ags and that this response was dependent on the endocytic receptor CD91 [51]. We then showed that a similar but class Ia-independent representation of hsp chaperoned antigens was involved in the case of the anti-15/0 tumor immune response [50]. Accordingly, LG-6 pMac exposed to tumor-derived gp96 and adoptively transferred into LG-6 hosts markedly impaired the growth of transplanted 15/0 tumor in a CD91-dependent manner.

In the case of hsp70, we went further to distinguish the respective role of the inducible hsp72 and the cognate or constitutively expressed hsc73. Although these two types of cytosolic hsp70 share very similar primary structure, they exhibit significant differences in their peptide- or ligand-binding domains, subcellular localization, and some of their function [52]. To be able to examine the tumor immunogenicity of each hsp70 isoform, we produced *X. laevis* recombinant cognate hsc73 and the inducible hsp72 from stable 15/0 tumor transfectants. Both hsp72 and hsc73-Ag complexes exhibited a similar ability for eliciting class Ia-mediated T-cell responses against minor H-Ag skin grafts. In contrast, our in vivo representation assay revealed that hsp72 was more potent than hsc73 in generating protective immune responses against the class Ia-negative 15/0 tumors in an Ag-dependent and putatively class Ib-mediated manner. This study provided the first evidence that although hsc73 is as potent as hsp72 in facilitating class Ia-restricted T-cell responses, it is less efficient than hsp72 in eliciting class Ia-unrestricted antitumor T-cell responses that are class Ib-mediated.

## 8.5 Conserved Roles of Nonclassical MHC and Innate T Cells in Tumor Immunity

As a method of immune evasion, tumors often downregulate their class Ia expression and thus facilitate their escape from conventional T-cell-mediated immune recognition and killing [53]. Importantly, loss of class Ia expression constitutes a loss of "self-signal" and can subsequently render malignant cells more susceptible to NK cell-mediated cytotoxicity. Consequentially, in order to avoid NK-mediated killing, many different types of tumors induce or upregulate the expression of class Ib genes [16]. Accordingly, an increased expression of certain class Ib molecules has been postulated to be an indicator of malignancy and/or intracellular stress [16]. Although the critical implication of classical MHC class Ia in tumor immune surveillance by eliciting effective antitumor CD8 cytotoxic T-cell effectors is well established from *Xenopus* to mammals, the roles of nonclassical MHC class Ib molecules and the effectors interacting with these molecules from NK to unconventional and innate T cells are less well understood.

The functional relevance of class Ib molecules in the cancer field is still unclear and often contradictory. Clinical studies have confirmed class Ib upregulated expression as a hallmark of certain tumors and shown that this typically correlates with unfavorable prognostics. HLA-E and HLA-G, in particular, have been shown to be indicators of poor clinical outcome in several different types of cancer [54–58]. On the other hand, other class Ib proteins, both in human and mouse, have been credited with the ability to mediate protective immunity against a variety of different cancers. In fact, due to their critical regulatory roles in immunity, certain class Ib molecules have emerged as attractive therapeutic targets against malignant neoplastic growths [59, 60]. Among potential class Ib targets, CD1d is perhaps the most studied. CD1d is critical for the development and function of CD1d-restricted invariant natural killer T-cells (iNKT) cells, which despite their relatively small numbers play critical regulatory roles promoting antitumor responses [59–61]. Several ongoing clinical trials are evaluating the effect of CD1d-mediated stimulation of iNKT cells with α-galactosylceramide (α-GalCer) on cancer patients (reviewed in [62]). Even though no clear tumor regression was observed, the iNKT-based therapies increased INF-γ blood levels, provided disease stabilization, and prolonged mean survival in patients no longer responding to chemo- or radiotherapies.

However, efficient clinical implementation of CD1 and iNKT cell-based therapies is still far from realization and requires a deeper and comprehensive understanding of the biology of this system.

From an evolutionary perspective, both class Ia and class Ib genes have been found in all jawed vertebrates studied to date (reviewed in [63]). Although relationships between evolutionarily distant class Ib molecules are difficult to establish, functional analogs, such as the primate HLA-E and the mouse Qa-1b, have been identified [64]. Representatives of the CD1 family of genes are found in mammals [65, 66], birds [67, 68], and reptiles [69] but in neither fish nor amphibians. In *X. laevis* there are at least 23 class Ib (*XNCs*) genes that, like other vertebrate class Ibs, are heterogeneous, less polymorphic, and less ubiquitously expressed than class Ia [39, 70–72]. Many of these *XNC* genes have an unusually high degree of conservation between *X. laevis* and *X. tropicalis* species both in primary sequence and genomic organization [70, 72]. The strong gene selection maintained in these two *Xenopus* species that diverged from a common ancestor as long ago as primates and rodents (~65 million years; [73]), is in support of important biological functions of *XNC* genes.

In this context, the high expression levels of several *XNC* genes by tumor lines derived from several independent lymphoid thymic tumors take on particular relevance. The possible involvement of certain *XNC* genes and XNC-restricted innate T cells in tumorigenesis and antitumor immunity in connection with hsps are all exciting avenues of investigation offered by the *Xenopus* model. To begin elucidating the functions of these *XNCs* in our tumor immunity model, we have chosen a loss-of-function reverse genetic approach based on RNA interference to silence *XNCs* at the level of the tumor. More specifically, the relevance of these *XNCs* for 15/0 tumorigenicity was investigated both indirectly by silencing *b2m*, which is usually required for surface expression of MHC class I molecules including class Ibs, and directly by silencing the expression of multiple *XNC* genes by targeting a consensus sequence shared by most *XNC* transcripts [74]. In fact in the case of *XNC10*, we were able to show the requirement of *b2m* surface expression. Interestingly, both types of silencing resulted in comparable results. 15/0 tumor transfectants deficient in either *b2m* or *XNCs* expression were more susceptible to NK-mediated killing but more resistant to killing by CD8 T cells in vitro. Moreover, 15/0 tumor transfectants were more tumorigenic in vivo upon transplantation in LG-15 adult recipients [74]. The faster tumor development of these *XNC*- or *b2m*-deficient tumor transfectants despite their decreased resistance to NK cell killing in vitro further suggested an important involvement of unconventional T cells interacting with XNC molecules rather than being restricted by MHC class Ia molecules.

However, further elucidation of the role of distinctive *XNC* gene products in this tumor model has revealed this to be more complex than previously thought. XNC10 represented an ideal candidate to focus on, since it is among the highest *XNC* expressed in 15/0 tumor and it is conserved, not only in *X. laevis* and *tropicalis* but also across ten different *Xenopus* species. Intriguingly, the specific silencing of *XNC10* in 15/0 tumor resulted in an acute rejection of these tumor transfectants by syngeneic LG-15 adults as well as naturally class Ia-deficient LG-15 tadpoles [75]. In tadpoles, the rejection was more potent toward 15/0 tumor transfectants with stronger *XNC10* knockdown. Furthermore, the rejection of XNC10-deficient tumors implicated cell-mediated cytotoxicity that could be enhanced by priming [75]. As such, *XNC10* is necessary for the immune evasion of the thymic-derived 15/0 tumors to escape immune recognition and class Ia-independent cytotoxicity. Taken together these findings suggest that various XNC molecules have different and possibly even opposing roles in immune surveillance, underlining the critical roles of class Ib molecules in tumor immunity. It is possible that different XNCs interact with distinct effector cells resulting in a balance between inhibitory and activating signals leading to either increased or decreased tumorigenicity.

## 8.6 Conserved Roles of Class Ib-Restricted Innate T Cell in Antitumor Immunity

Among MHC class Ib-restricted effector cells, innate T (iT) cells such as CD1d-restricted iNKT cells have recently emerged as a potentially critical component of tumor immunity as they can orchestrate both innate and adaptive immunity [76–79]. These lymphocytes are T cells with natural killer cell markers and expressing semi-invariant T-cell receptor (TCR) repertoires [14]. Although iT cells generally occur at low frequencies [80], they can control immune responses via rapid and potent release of either pro-inflammatory or anti-inflammatory cytokines [81].

Notably, we have recently demonstrated that iT cells are not only conserved in *Xenopus*, but may constitute a more prominent component of their immune system than in mammals, especially during tadpole life [82]. To date we have been able to characterize the iT cell subset restricted by XNC10 [15, 82]. Using a reverse genetic approach combining transgenesis with RNA interference, we showed that XNC10 is required for the development of these iT cells. Furthermore, based on TCR diversity, XNC10 tetramer binding, and CD8 antibody staining, two subpopulations have been characterized within the *Xenopus* XNC10-restricted iT cells, type I XNC10-$T^{+}$/CD8− and XNC10-$T^{dim+}$CD8$^{dim+}$, which are reminiscent of mammalian type I iNKT and type II NKT cells, respectively [82].

Interestingly, rapid infiltration of XNC10-iT cells is observed following intraperitoneal 15/0 tumor transplantation into LG-15 tadpoles [75]. Similar early infiltration of XNC10-iT cells also occurs when transplanting ff-2 tumor into inbred F tadpoles (Banach and Robert, unpublished observations). Intriguingly, knock-down of XNC10 in 15/0 tumor triggers a substantially increased infiltration of XNC10-iT cells, which is again consistent with the use of XNC10 as an immune evasion strategy by the 15/0 tumors.

## 8.7 Conclusions and Perspective

Antigen presentation by classical MHC class Ia molecules as a way to induce potent antigen-specific CD8 T-cell responses is a pivotal component of the immune surveillance system. More specifically, in the context of tumor immune surveillance, APCs are postulated to acquire tumor antigens generated by deregulated gene expression and/or mutations from the malignant cell and then generate an adaptive T-cell response specific to these antigens. Hsps, such as cytosolic HSP70, and ER-resident gp96 can contribute to elicit this antitumor response by chaperoning tumor antigens thus facilitating efficient cross-presentation as well as by enhancing the co-stimulation responses important for potent activation of T cells.

Here, we propose that hsps, classical MHC class Ia, nonclassical MHC class Ib molecules, and their respective effector cells are integrated in an ancestral immune surveillance system (Fig. 8.2). Indeed, the critical involvement of class Ib molecules

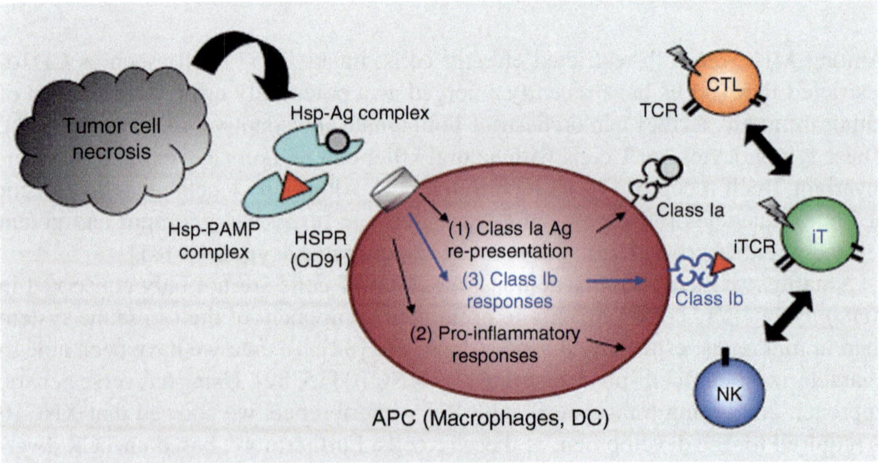


**Fig. 8.2** Proposed ancestral immune surveillance system. Hsp-peptide complexes released in the extracellular compartment from infected or stressed cells (e.g., apoptosis, cell lysis) are internalized by APCs through receptor-mediated endocytosis (e.g., CD91). (1) Antigenic peptides channeled into the class Ia presentation pathway activate CD8 T cells. (2) Hsps internalized by the same receptors or interacting with other receptors (e.g., TLRs) stimulate pro-inflammatory responses. (3) Hsps are proposed to also stimulate class Ib-mediated responses by an as yet unknown mechanism that is likely to be Ag-specific and involve iT cell populations

in amphibian hsp-mediated antitumor responses and the finding that class Ib-restricted antitumor iT cells are present and prominent outside mammals raise the intriguing possibility that this system is ancestral and widespread across jawed vertebrates. Although the role of nonclassical MHC molecules and unconventional T cells, including iT cells in tumor immunity, is still far from fully elucidated, the inherent ability of class Ib molecules to present nonprotein antigens such as lipids and other conserved molecular motifs or patterns offers an extended avenue of detectable antitumor determinants. The limited variation of these class Ib-binding patterns and their conservation during evolution could be exploited as target of choice for future immunotherapy. In addition, the potent and rapid activation of unconventional class Ia-unrestricted T cells such as iT cells may be critical in promoting antitumor versus pro-tumor suppressive microenvironments.

In this context, the ability of hsps to also promote iT cell responses through class Ib molecules is a promising new avenue to investigate. Given that during tumor progression class Ia molecules are often downregulated, cancer immunotherapies that exploit class Ia-restricted T-cell effectors are usually insufficient to maintain potent antitumor responses. Conversely, as some class Ib molecules remain expressed on tumors or in some cases are even upregulated, these molecules and their interacting immune effector cells could serve as additional persisting immunogenic targets. Thus, the elucidation of the roles of class Ib molecules in tumor immunity is of fundamental scientific and clinical interest.

**Acknowledgments** We would like to thank Dr. Edith Lord for critical reading of the manuscript. This work was supported by a R24-AI-059830 grant from the National Institute of Allergy and Infectious Diseases (NIH/NIAID) and from the Kesel Fund of Rochester Area Community Foundation, Rochester, NY. M.B. was supported by a predoctoral fellowship Ruth L. Kirschstein Predoctoral F31 (F31CA192664) from the National Cancer Institute (NIH/NCI). E-S.E. was supported by the National Science Foundation IOS-1456213 and a 2015 Career in Immunology Fellowship from the American Association of Immunologists.

## References

1. Hendrick JP, Hartl FU (1993) Molecular chaperone functions of heat-shock proteins. Annu Rev Biochem 62:349–384
2. Calderwood SK, Gong J (2016) Heat shock proteins promote cancer: it's a protection racket. Trends Biochem Sci 41:311
3. Srivastava P (2002a) Interaction of heat shock proteins with peptides and antigen presenting cells: chaperoning of the innate and adaptive immune responses. Annu Rev Immunol 20:395–425
4. Srivastava P (2002b) Roles of heat-shock proteins in innate and adaptive immunity. Nat Rev Immunol 2:185–194
5. Bendz H, Ruhland SC, Pandya MJ, Hainzl O, Riegelsberger S, Brauchle C, Mayer MP, Buchner J, Issels RD, Noessner E (2007) Human heat shock protein 70 enhances tumor antigen presentation through complex formation and intracellular antigen delivery without innate immune signaling. J Biol Chem 282:31688–31702

6. Blachere NE, Li Z, Chandawarkar RY, Suto R, Jaikaria NS, Basu S, Udono H, Srivastava PK (1997) Heat shock protein-peptide complexes, reconstituted in vitro, elicit peptide-specific cytotoxic T lymphocyte response and tumor immunity. J Exp Med 186:1315–1322
7. Suto R, Srivastava PK (1995) A mechanism for the specific immunogenicity of heat shock protein-chaperoned peptides. Science 269:1585–1588
8. Robert J, Gantress J, Rau L, Bell A, Cohen N (2002) Minor histocompatibility antigen-specific MHC-restricted CD8 T cell responses elicited by heat shock proteins. J Immunol 168:1697–1703
9. Robert J, Menoret A, Basu S, Cohen N, Srivastava PR (2001) Phylogenetic conservation of the molecular and immunological properties of the chaperones gp96 and hsp70. Eur J Immunol 31:186–195
10. Basu S, Binder RJ, Ramalingam T, Srivastava PK (2001) CD91 is a common receptor for heat shock proteins gp96, hsp90, hsp70, and calreticulin. Immunity 14:303–313
11. Binder RJ, Han DK, Srivastava PK (2000) CD91: a receptor for heat shock protein gp96. Nat Immunol 1:151–155
12. Binder RJ, Srivastava PK (2005) Peptides chaperoned by heat-shock proteins are a necessary and sufficient source of antigen in the cross-priming of CD8+ T cells. Nat Immunol 6:593–599
13. Lammert E, Arnold D, Nijenhuis M, Momburg F, Hammerling GJ, Brunner J, Stevanovic S, Rammensee HG, Schild H (1997) The endoplasmic reticulum-resident stress protein gp96 binds peptides translocated by TAP. Eur J Immunol 27:923–927
14. Adams EJ, Luoma AM (2013) The adaptable major histocompatibility complex (MHC) fold: structure and function of nonclassical and MHC class I-like molecules. Annu Rev Immunol 31:529–561
15. Edholm ES, Grayfer L, Robert J (2014b) Evolution of nonclassical MHC-dependent invariant T cells. Cell Mol Life Sci 71:4763–4780
16. Gleimer M, Parham P (2003) Stress management: MHC class I and class I-like molecules as reporters of cellular stress. Immunity 19:469–477
17. Gomes AQ, Correia DV, Silva-Santos B (2007) Non-classical major histocompatibility complex proteins as determinants of tumour immunosurveillance. EMBO Rep 8:1024–1030
18. Robert J, Goyos A, Nedelkovska H (2009) Xenopus, a unique comparative model to explore the role of certain heat shock proteins and non-classical MHC class Ib gene products in immune surveillance. Immunol Res 45:114
19. Robert J, Ohta Y (2009) Comparative and developmental study of the immune system in Xenopus. Dev Dyn 238:1249–1270
20. Flajnik MF, Du Pasquier L (1990) The major histocompatibility complex of frogs. Immunol Rev 113:47–63
21. Flajnik MF, Kaufman JF, Hsu E, Manes M, Parisot R, Du Pasquier L (1986) Major histocompatibility complex-encoded class I molecules are absent in immunologically competent Xenopus before metamorphosis. J Immunol 137:3891–3899
22. Kau CL, Turpen JB (1983) Dual contribution of embryonic ventral blood island and dorsal lateral plate mesoderm during ontogeny of hemopoietic cells in Xenopus laevis. J Immunol 131:2262–2266
23. Du Pasquier L, Weiss N (1973) The thymus during the ontogeny of the toad Xenopus laevis: growth, membrane-bound immunoglobulins and mixed lymphocyte reaction. Eur J Immunol 3:773–777
24. Bechtold TE, Smith PB, Turpen JB (1992) Differential stem cell contributions to thymocyte succession during development of Xenopus laevis. J Immunol 148:2975–2982
25. Turpen JB, Smith PB (1989) Precursor immigration and thymocyte succession during larval development and metamorphosis in Xenopus. J Immunol 142:41–47
26. Flajnik MF, Du Pasquier L (1988) MHC class I antigens as surface markers of adult erythrocytes during the metamorphosis of Xenopus. Dev Biol 128:198–206
27. Rollins-Smith LA, Flajnik MF, Blair PJ, Davis AT, Green WF (1997) Involvement of thyroid hormones in the expression of MHC class I antigens during ontogeny in Xenopus. Dev Immunol 5:133–144

28. Salter-Cid L, Nonaka M, Flajnik MF (1998) Expression of MHC class Ia and class Ib during ontogeny: high expression in epithelia and coregulation of class Ia and lmp7 genes. J Immunol 160:2853–2861
29. Du Pasquier L, Schwager J, Flajnik MF (1989) The immune system of Xenopus. Annu Rev Immunol 7:251–275
30. Goyos A, Robert J (2009) Tumorigenesis and anti-tumor immune responses in Xenopus. Front Biosci 14:167–176
31. Du Pasquier L, Robert J (1992) In vitro growth of thymic tumor cell lines from Xenopus. Dev Immunol 2:295–307
32. Robert J, Guiet C, Du Pasquier L (1994) Lymphoid tumors of Xenopus laevis with different capacities for growth in larvae and adults. Dev Immunol 3:297–307
33. Earley EM, Reinschmidt DC, Tompkins R, Gebhardt BM (1995) Tissue culture of a mixed cell thymic tumor from Xenopus laevis. In Vitro Cell Dev Biol Anim 31:255–257
34. Du Pasquier L, Wilson M, Sammut B (2009) The fate of duplicated immunity genes in the dodecaploid Xenopus ruwenzoriensis. Front Biosci 14:177–191
35. Robert J, Guiet C, Du Pasquier L (1995) Ontogeny of the alloimmune response against a transplanted tumor in Xenopus laevis. Differentiation 59:135–144
36. Du Pasquier L, Courtet M, Robert J (1995) A Xenopus lymphoid tumor cell line with complete Ig genes rearrangements and T-cell characteristics. Mol Immunol 32:583–593
37. Chretien I, Robert J, Marcuz A, Garcia-Sanz JA, Courtet M, Du Pasquier L (1996) CTX, a novel molecule specifically expressed on the surface of cortical thymocytes in Xenopus. Eur J Immunol 26:780–791
38. Robert J, Cohen N (1999) In vitro differentiation of a CD4/CD8 double-positive equivalent thymocyte subset in adult Xenopus. Int Immunol 11:499–508
39. Goyos A, Ohta Y, Guselnikov S, Robert J (2009) Novel nonclassical MHC class Ib genes associated with CD8 T cell development and thymic tumors. Mol Immunol 46:1775–1786
40. Robert J, Guiet C, Cohen N, Du Pasquier L (1997) Effects of thymectomy and tolerance induction on tumor immunity in adult Xenopus laevis. Int J Cancer 70:330–334
41. Goyos A, Cohen N, Gantress J, Robert J (2004) Anti-tumor MHC class Ia-unrestricted CD8 T cell cytotoxicity elicited by the heat shock protein gp96. Eur J Immunol 34:2449–2458
42. Horton TL, Minter R, Stewart R, Ritchie P, Watson MD, Horton JD (2000) Xenopus NK cells identified by novel monoclonal antibodies. Eur J Immunol 30:604–613
43. Rau L, Gantress J, Bell A, Stewart R, Horton T, Cohen N, Horton J, Robert J (2002) Identification and characterization of Xenopus CD8+ T cells expressing an NK cell-associated molecule. Eur J Immunol 32:1574–1583
44. Horton TL, Stewart R, Cohen N, Rau L, Ritchie P, Watson MD, Robert J, Horton JD (2003) Ontogeny of Xenopus NK cells in the absence of MHC class I antigens. Dev Comp Immunol 27:715–726
45. Delneste Y, Magistrelli G, Gauchat J, Haeuw J, Aubry J, Nakamura K, Kawakami-Honda N, Goetsch L, Sawamura T, Bonnefoy J, Jeannin P (2002) Involvement of LOX-1 in dendritic cell-mediated antigen cross-presentation. Immunity 17:353–362
46. Facciponte JG, Wang XY, Subjeck JR (2007) Hsp110 and Grp170, members of the Hsp70 superfamily, bind to scavenger receptor-A and scavenger receptor expressed by endothelial cells-I. Eur J Immunol 37:2268–2279
47. Murshid A, Borges TJ, Calderwood SK (2015) Emerging roles for scavenger receptor SREC-I in immunity. Cytokine 75:256–260
48. Asea A, Rehli M, Kabingu E, Boch JA, Bare O, Auron PE, Stevenson MA, Calderwood SK (2002) Novel signal transduction pathway utilized by extracellular HSP70: role of toll-like receptor (TLR) 2 and TLR4. J Biol Chem 277:15028–15034
49. Warger T, Hilf N, Rechtsteiner G, Haselmayer P, Carrick DM, Jonuleit H, von Landenberg P, Rammensee HG, Nicchitta CV, Radsak MP, Schild H (2006) Interaction of TLR2 and TLR4 ligands with the N-terminal domain of Gp96 amplifies innate and adaptive immune responses. J Biol Chem 281:22545–22553

50. Nedelkovska H, Robert J (2013) Hsp72 mediates stronger antigen-dependent non-classical MHC class Ib anti-tumor responses than hsc73 in Xenopus laevis. Cancer Immun 13:4
51. Robert J, Ramanayake T, Maniero GD, Morales H, Chida AS (2008) Phylogenetic conservation of glycoprotein 96 ability to interact with CD91 and facilitate antigen cross-presentation. J Immunol 180:3176–3182
52. Callahan MK, Chaillot D, Jacquin C, Clark PR, Menoret A (2002) Differential acquisition of antigenic peptides by Hsp70 and Hsc70 under oxidative conditions. J Biol Chem 277:33604–33609
53. Zitvogel L, Tesniere A, Kroemer G (2006) Cancer despite immunosurveillance: immunoselection and immunosubversion. Nat Rev Immunol 6:715–727
54. Benevolo M, Mottolese M, Tremante E, Rollo F, Diodoro MG, Ercolani C, Sperduti I, Lo Monaco E, Cosimelli M, Giacomini P (2011) High expression of HLA-E in colorectal carcinoma is associated with a favorable prognosis. J Transl Med 9:184
55. de Kruijf EM, Sajet A, van Nes JG, Natanov R, Putter H, Smit VT, Liefers GJ, van den Elsen PJ, van de Velde CJ, Kuppen PJ (2010) HLA-E and HLA-G expression in classical HLA class I-negative tumors is of prognostic value for clinical outcome of early breast cancer patients. J Immunol 185:7452–7459
56. He X, Dong DD, Yie SM, Yang H, Cao M, Ye SR, Li K, Liu J, Chen J (2010) HLA-G expression in human breast cancer: implications for diagnosis and prognosis, and effect on allocytotoxic lymphocyte response after hormone treatment in vitro. Ann Surg Oncol 17:1459–1469
57. Ye SR, Yang H, Li K, Dong DD, Lin XM, Yie SM (2007) Human leukocyte antigen G expression: as a significant prognostic indicator for patients with colorectal cancer. Mod Pathol 20:375–383
58. Yie SM, Yang H, Ye SR, Li K, Dong DD, Lin XM (2007) Expression of HLA-G is associated with prognosis in esophageal squamous cell carcinoma. Am J Clin Pathol 128:1002–1009
59. McEwen-Smith RM, Salio M, Cerundolo V (2015) The regulatory role of invariant NKT cells in tumor immunity. Cancer Immunol Res 3:425–435
60. Robertson FC, Berzofsky JA, Terabe M (2014) NKT cell networks in the regulation of tumor immunity. Front Immunol 5:543
61. Nagato K, Motohashi S, Ishibashi F, Okita K, Yamasaki K, Moriya Y, Hoshino H, Yoshida S, Hanaoka H, Fujii S, Taniguchi M, Yoshino I, Nakayama T (2012) Accumulation of activated invariant natural killer T cells in the tumor microenvironment after alpha-galactosylceramide-pulsed antigen presenting cells. J Clin Immunol 32:1071–1081
62. Altman JB, Benavides AD, Das R, Bassiri H (2015) Antitumor responses of invariant natural killer T cells. J Immunol Res 2015:652875
63. Flajnik MF, Kasahara M (2001) Comparative genomics of the MHC: glimpses into the evolution of the adaptive immune system. Immunity 15:351–362
64. Yeager M, Kumar S, Hughes AL (1997) Sequence convergence in the peptide-binding region of primate and rodent MHC class Ib molecules. Mol Biol Evol 14:1035–1041
65. Brossay L, Chioda M, Burdin N, Koezuka Y, Casorati G, Dellabona P, Kronenberg M (1998) CD1d-mediated recognition of an alpha-galactosylceramide by natural killer T cells is highly conserved through mammalian evolution. J Exp Med 188:1521–1528
66. Dascher CC (2007) Evolutionary biology of CD1. Curr Top Microbiol Immunol 314:3–26
67. Miller MM, Wang C, Parisini E, Coletta RD, Goto RM, Lee SY, Barral DC, Townes M, Roura-Mir C, Ford HL, Brenner MB, Dascher CC (2005) Characterization of two avian MHC-like genes reveals an ancient origin of the CD1 family. Proc Natl Acad Sci U S A 102:8674–8679
68. Salomonsen J, Sorensen MR, Marston DA, Rogers SL, Collen T, van Hateren A, Smith AL, Beal RK, Skjodt K, Kaufman J (2005) Two CD1 genes map to the chicken MHC, indicating that CD1 genes are ancient and likely to have been present in the primordial MHC. Proc Natl Acad Sci U S A 102:8668–8673
69. Yang Z, Wang C, Wang T, Bai J, Zhao Y, Liu X, Ma Q, Wu X, Guo Y, Zhao Y, Ren L (2015) Analysis of the reptile CD1 genes: evolutionary implications. Immunogenetics 67:337–346

70. Edholm ES, Goyos A, Taran J, De Jesus Andino F, Ohta Y, Robert J (2014a) Unusual evolutionary conservation and further species-specific adaptations of a large family of nonclassical MHC class Ib genes across different degrees of genome ploidy in the amphibian subfamily Xenopodinae. Immunogenetics 66:411–426
71. Flajnik MF, Kasahara M, Shum BP, Salter-Cid L, Taylor E, Du Pasquier L (1993) A novel type of class I gene organization in vertebrates: a large family of non-MHC-linked class I genes is expressed at the RNA level in the amphibian Xenopus. EMBO J 12:4385–4396
72. Goyos A, Sowa J, Ohta Y, Robert J (2011) Remarkable conservation of distinct nonclassical MHC class I lineages in divergent amphibian species. J Immunol 186:372–381
73. Evans BJ (2008) Genome evolution and speciation genetics of clawed frogs (Xenopus and Silurana). Front Biosci 13:4687–4706
74. Goyos A, Guselnikov S, Chida AS, Sniderhan LF, Maggirwar SB, Nedelkovska H, Robert J (2007) Involvement of nonclassical MHC class Ib molecules in heat shock protein-mediated anti-tumor responses. Eur J Immunol 37:1494–1501
75. Haynes-Gilmore N, Banach M, Edholm ES, Lord E, Robert J (2014) A critical role of nonclassical MHC in tumor immune evasion in the amphibian Xenopus model. Carcinogenesis 35:1807–1813
76. Bassiri H, Das R, Guan P, Barrett DM, Brennan PJ, Banerjee PP, Wiener SJ, Orange JS, Brenner MB, Grupp SA, Nichols KE (2014) iNKT cell cytotoxic responses control T-lymphoma growth in vitro and in vivo. Cancer Immunol Res 2:59–69
77. Brennan PJ, Brigl M, Brenner MB (2013) Invariant natural killer T cells: an innate activation scheme linked to diverse effector functions. Nat Rev Immunol 13:101–117
78. Pilones KA, Aryankalayil J, Babb JS, Demaria S (2014) Invariant natural killer T cells regulate anti-tumor immunity by controlling the population of dendritic cells in tumor and draining lymph nodes. J Immunother Cancer 2:37
79. Pilones KA, Aryankalayil J, Demaria S (2012) Invariant NKT cells as novel targets for immunotherapy in solid tumors. Clin Dev Immunol 2012:720803
80. Kronenberg M (2005) Toward an understanding of NKT cell biology: progress and paradoxes. Annu Rev Immunol 23:877–900
81. Matsuda JL, Mallevaey T, Scott-Browne J, Gapin L (2008) CD1d-restricted iNKT cells, the 'Swiss-Army knife' of the immune system. Curr Opin Immunol 20(3):358–368. https://doi.org/10.1016/j.coi.2008.03.018. Epub 2008 May 22
82. Edholm ES, Albertorio Saez LM, Gill AL, Gill SR, Grayfer L, Haynes N, Myers JR, Robert J (2013) Nonclassical MHC class I-dependent invariant T cells are evolutionarily conserved and prominent from early development in amphibians. Proc Natl Acad Sci U S A 110:14342–14347

# Chapter 9
# Inhibition of HSPs for Enhanced Immunity

**Ronald J. Fecek, Subhara Raveendran, Manoj Chelvanambi, and Walter J. Storkus**

**Abstract** Heat shock proteins (HSPs) are highly abundant proteins found in all cell types in the body, where they comprise approximately 1–2% of the cellular proteome. Due to the physiologically stressful conditions of the progressive tumor microenvironment (TME, i.e., hypoxia, acidosis, and high interstitial fluid pressure), expression of HSPs in tumor cells can be increased by a factor of two- to tenfold over that found in normal cells. Larger HSPs (HSP70 and HSP90) maintain the structural integrity of a broad range of tumor client proteins associated with oncogenesis and disease progression. HSPs can also be translocated to the tumor cell surface or shed into the extracellular space where they have recently been found to serve as "chaperokines" capable of modulating the function of antigen-presenting cells and immune effector cells. This chapter will provide a summary of the pleiotropic impact of HSPs on tumor immunity and suggest strategies by which

R. J. Fecek
Department of Dermatology, University of Pittsburgh School of Medicine, Pittsburgh, PA, USA

S. Raveendran
Department of Human Genetics, University of Pittsburgh School of Medicine, Pittsburgh, PA, USA

M. Chelvanambi
Department of Immunology, University of Pittsburgh School of Medicine, Pittsburgh, PA, USA

W. J. Storkus (✉)
Department of Dermatology, University of Pittsburgh School of Medicine, Pittsburgh, PA, USA

Department of Immunology, University of Pittsburgh School of Medicine, Pittsburgh, PA, USA

Department of Pathology, University of Pittsburgh School of Medicine, Pittsburgh, PA, USA

Department of Bioengineering, University of Pittsburgh School of Medicine, Pittsburgh, PA, USA

University of Pittsburgh Graduate School of Public Health, Pittsburgh, PA, USA

Melanoma Program of the University of Pittsburgh Cancer Institute, Pittsburgh, PA, USA
e-mail: storkuswj@upmc.edu



R. J. Binder, P. K. Srivastava (eds.), *Heat Shock Proteins in the Immune System*,
https://doi.org/10.1007/978-3-319-69042-1_9

HSP inhibitors (HSPi) might be best applied to optimize the antitumor efficacy of combination immunotherapy approaches.

**Keywords** Adoptive cellular therapy · Cytotoxic T lymphocytes · Heat shock protein · Histone deacetylase inhibitors · HSP inhibitors · Immunotherapy · Regulatory cells · Tumor microenvironment · Vaccine · Vascular normalization

## 9.1 Tumor Cell Overexpression of HSPs Is Correlated with Poor Clinical Outcome

All major HSP families (HSP27, HSP40, HSP60, HSP70, HSP90, HSP110), a range of co-chaperone molecules (such as BAG3 among others), as well as drivers of HSP transcription (such as HSF1) have been reported to be activated/overexpressed in many forms of human cancer, where they may serve as prognostic biomarkers of poor clinical outcome and resistance to conventional chemotherapy and/or radiotherapy [1–13]. Higher expression levels of tumor-associated HSF1 and HSPs have been associated with tumor cell proliferative potential, survival and invasiveness/aggressiveness [2, 7, 14], disease stage [15], overall survival (OS; [15–20]), and progression-free survival (PFS; [5, 9, 21–23]) in human acute myelogenous leukemia (AML), osteosarcomas, and carcinomas of the bladder, breast, cervix, colon endometrium, liver, lung, prostate, stomach, and uterus. In hepatocellular carcinoma (HCC), the tumor suppressor molecule HBP21, which limits the interaction of HSP70 with its client proteins, is downregulated in association with advanced clinical stage and poor prognosis [14]. Interestingly, a loss-of-function mutation in HSP110 (HSP110ΔE9) has been identified in colorectal carcinoma (CRC), leading to limited translocation of wild-type HSP110 to the tumor cell nucleus, and patients with this mutation exhibit increased tumor sensitivity to genotoxic chemotherapeutic agents [9]. Similarly, CRC patients with microsatellite instability and deletions in HSP110 display superior clinical responses to treatment with chemotherapy [24]. Transcriptional activation of HSF1 in tumor cells may occur via the direct action of oncogenes/oncogenic signaling pathways [25–27], leading to facilitated nuclear translocation of HSF1 and subsequent transcription of HSPs (i.e., HSP 27, HSP70, HSP90; [7, 28]) based on hypoxia response elements (HREs) in their promoter regions [29]. Depletion of HSF1 leads to reduced tumor cell expression of HSPs (i.e., HSP27, HSP40, HSP70, and HSP90) and to tumor cell apoptosis and slowed tumor growth in vivo [30].

## 9.2 Impact of Tumor Cell Intrinsic HSPs on Cancer Cell Development, Progression, and Interaction with the Immune System

HSP90 together with HSP40, HSP70, HIP, and HOP forms a super-chaperone machine that operationally refolds and stabilizes a broad range of denatured tumor-associated client proteins in an ATP-dependent manner [31]. This effectively extends the functional life span of HSP client proteins, many of which are known to play key roles in support of the molecular "hallmarks of cancer" (i.e., pro-oncogenic pathways associated with tumor development, survival (cytoprotection), progression and metastasis, and resistance to immune-mediated rejection; [32–34]; Fig. 9.1).

In the latter instance, tumor cell HSP90 overexpression has profound impact on the inactivation of protective host immune responses. For example, the HSP90 client protein HER2 can be translocated into nucleus of human breast carcinoma cells where it can bind and transactivate the COX-2 promoter in support of the production of prostaglandin E2 ($PGE_2$; [36]). Similarly, the activity of tumor cytosolic prostaglandin E2 synthase is regulated by HSP90 [37, 38]. Notably, $PGE_2$ drives the induction and maintenance of cancer-associated myeloid-derived suppressor cells (MDSC), the enhancement of Treg functionality, and suppression of antitumor

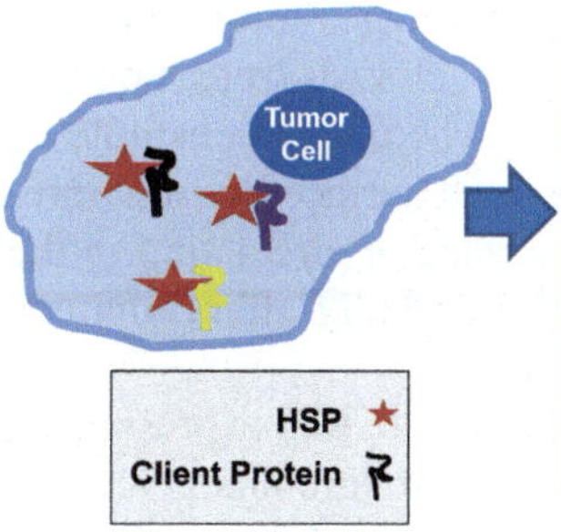


| Cancer Hallmark | HSP(90) Client Examples |
|---|---|
| Aerobic Glycolysis | GLUT1, SDHB |
| Enhanced Proliferation | BRAF, CRAF, EGFR, Her2, IGF-1R, KIT, RAF-1, SRC |
| Evading Cell Death | AKT, BCL2, BCLxL, cIAP1, MCL1, PTEN, Survivin |
| Evading Growth Suppression | β-Catenin, CCND1, CDK4, E2F1, p16, p53, RB |
| Genomic Instability | ATM, CHK1, DNA-PK, MGMT, RAD51 |
| Pro-Angiogenesis | HIF-1α, PDGFR, STAT3, VEGFR1, VEGFR2 |
| Pro-Invasion/Metastasis | ANGPTL4, EPHA2, FAK, LOX, MET, MMP2, MMP9 |
| Replicative Immortality | ALDH1, OCT4, NANOG, TERT |
| Carcinogenic Inflammation | FGFR, NLRs, RIP, SGTI, TLRs |
| Immune Evasion | HER2, MUC1, NOS2, STAT3 |

**Fig. 9.1** Tumor HSPs reinforce the hallmarks of cancer. Hanahan and Weinberg (Cell. 2011;144:646-74, Cell. 2000;100:57–70 [35]) defined hallmarks of cancer that support tumor development, proliferation, growth, invasiveness/metastasis, and resistance to programmed cell death and immune-mediated control. Many proteins that are operationally involved in these pro-oncogenic programs represent HSP(90) client proteins (with some relevant examples provided for each cancer hallmark)

T/NK effector cell functions [39, 40]. HSP90 client cancer-associated MUC1 mucin inhibits human T-cell activation and proliferation [41, 42]. Inducible nitric oxide synthase (NOS2) is also a tumor cell overexpressed HSP90 client (https://www.picard.ch/downloads/Hsp90interactors.pdf) that has been correlated with poor clinical prognosis in a range of cancer types [43–45]. Tumor NOS2 promotes the induction of functional myeloid-derived suppressor cells via the modulation of VEGF release [46, 47]. Furthermore, HSP90 client STAT3-dependent signaling in tumor cells promotes TGF-β production/secretion and the suppression of innate and adaptive immunity [48].

As will be discussed later in this chapter, HSPs also play central contributions to the function and orchestration of the antigen-processing and MHC-presentation machinery of tumor cells which directly impact the ability of these aberrant cell populations to be recognized and regulated by protective T effector cells [49].

As a consequence, targeted disruption of the expression or function in tumor cells would be expected to profoundly limit cancer growth, progression, and metastases and to have dichotomous effects on tumor cell regulation by the immune system. On one hand, the destabilization of HSP clients linked to the development of suppressor MDSC and Treg would be anticipated to improve chances for protective immune cell eradication of neoplastic cells, while on the other hand, if HSP function were too severely antagonized, $CD4^+$ and $CD8^+$ T-cell recognition of tumor cells might be hindered.

## 9.3 Immune Modulation by Tumor Extracellular HSPs (eHSPs)

In addition to their conventional molecular chaperone functions within tumor cells, it has become increasingly apparent that under certain conditions, tumor HSPs can access the extracellular microenvironment (i.e., eHSPs) via cell lysis (releasing free eHSPs), in the context of shed microvessels or exosomes where they constitute an abundant cargo, and via their translocation to the tumor cell surface during a pre-apoptotic state (Fig. 9.2). Once expressed on the cell surface or circulating in the tumor interstitial fluid or in serum/plasma, eHSP can profoundly affect immune function in either a positive or negative manner [50, 51].

eHSPs (i.e., HSP70, HSP90, gp96) along with calreticulin (CRT) expressed on the cell surface of dying/dead tumor cells are characteristic of "immunogenic cell death" (ICD), whereby eHSPs and CRT provide an "eat me" signal to host DC via their binding to HSP receptors (i.e., HSP-R, including LRP1/CD91, CD40, TLR2/4, CD14, CD35, Lox-1, and SR-A), leading to the uptake of $eHSP^+$ tumor cell bodies and the processing and (cross)presentation of antigens by matured DC that consequently sponsor the activation, expansion, and differentiation of antitumor T cells (Fig. 9.2; [52–56]). A range of cancer chemotherapy agents (including anthracyclines such as doxorubicin, the proteasome inhibitor bortezomib, irradiation, and

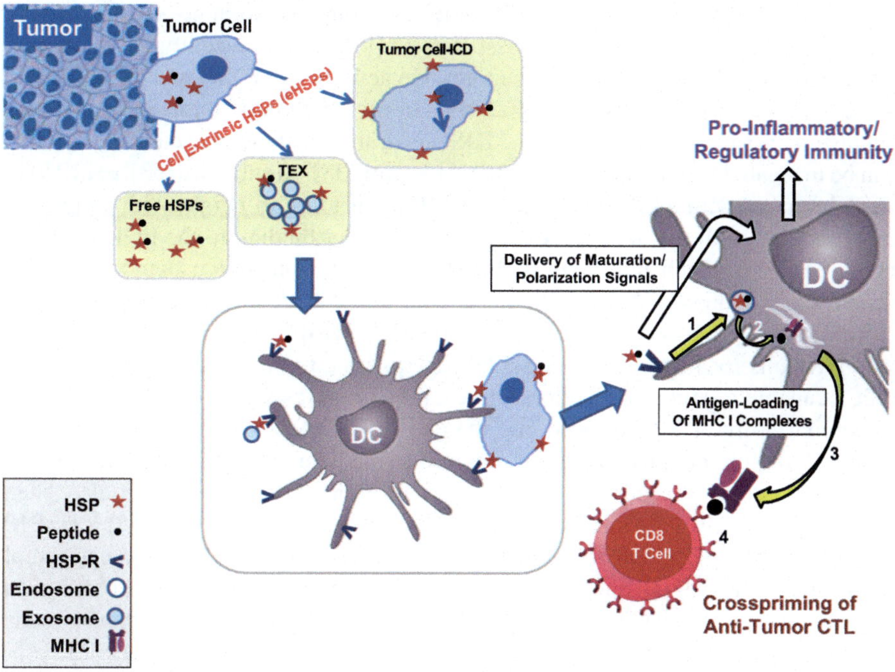


**Fig. 9.2** The multifactorial role of eHSPs in modulating adaptive immunity against tumor-associated antigens. Although HSPs, including HSP90, are well-appreciated for their capacity to sustain the life span of tumor-intrinsic proteins that are operationally involved in cancer development and progression, it has become increasingly evident that the immune system has evolved key sensors to monitor tumor-extrinsic HSPs (eHSPs released from dying tumor cells, shed/secreted in the context of subcellular exosomes or expressed on the tumor cell surface during the process of "immunogenic cell death (ICD)"). Tumor eHSP complexes (containing bound peptides or decorating the surface of tumor exosomes (TEX) or tumor cells undergoing ICD) can bind to receptors (such as CD91 or SREC-1, etc.) expressed by dendritic cells, leading to the transmission of activating or regulatory signals that promote DC maturation into inflammatory/regulatory antigen-(cross) presenting cells capable of priming/restimulating antitumor T cells (including cytotoxic T lymphocytes; CTL) or driving the expansion and functional activation of Treg. After HSP-peptide complexes bind to receptors on the DC cell surface, these complexes are endocytosed (1) and delivered to the endoplasmic reticulum (ER), where the HSP-cargo peptides may be loaded into nascent MHC class I complexes (2), making them transport-competent to the cell surface for subsequent recognition cognate class I-restricted CD8$^+$ T cells (3). Endocytosed HSP-peptide complexes may also access the MHC class II processing pathway, resulting in coordinate DC induction of tumor antigen-specific CD4$^+$ T cells (not shown)

photodynamic therapy (PDT)) have all been reported to increase tumor cell surface eHSP60, eHSP70, and eHSP90 expression and the process of ICD [55, 57–60]. eHSP$^+$ tumor cells have also been reported to serve as superior targets for NK cell-mediated killing [61].

While tumor cell-bound eHSPs appear to be largely immunostimulatory in their nature, cell-free tumor eHSPs (in the form of shed exosomes or soluble proteins) can lead to either the promotion or inhibition of protective immune cell function. In the case of tumor exosome (TEX)-associated eHSPs, it has recently been reported that eHSP70 in TEX enhances the suppressive activity of MDSC in a pSTAT3- and TLR2/MyD88-dependent manner [62, 63]. In mice treated with eHSP70, the percentage of peripheral $CD11b^{+}Gr\text{-}1^{+}$ MDSC is dramatically increased, although this can be mitigated by the injection of a blocking anti-HSP70 antibody [63]. eHSP70 in TEX interacts with TLR2 expressed by MDSC leading to heightened suppressor cell function, with the injection of a peptide aptamer binding to the ECD of membrane eHSP70 which interferes with eHSP70/TLR2 interaction demonstrated to decrease the number of MDSCs in tumor-bearing C57Bl/6 mice in association with slowed tumor progression [64]. Conversely, TEX eHSP70 has been shown to activate NK cells for improved tumoricidal function and to promote the maturation of APCs capable of stimulating alloreactive $CD4^{+}$ T-cell responses in vivo [65].

Similarly, free HSPs released into the extracellular space upon physical disruption of dying tumor cells have been reported to mediate dichotomous effects on the immune system. For instance, high doses of tumor-free HSPs have also been reported to inhibit the generation of antigen-specific T cells [50, 66–68]. eHSP10 can be overexpressed and released by ovarian carcinoma cells into the peripheral blood and ascites of patients, where it has been reported to reduce T-cell expression of a member of the TCR signaling complex (CD3-ζ), rendering T effector cells hyporesponsive to subsequent antigenic stimulation [69]. Tumor shed eHSP27 promotes monocyte differentiation into tumor-associated macrophages (TAM) expressing $MHCII^{low}CD86^{low}PD\text{-}L1^{high}ILT2^{high}ILT4^{high}$ phenotype, known to be pro-angiogenic in breast cancer patients [70]. Free eHSP60 promotes T effector cell cross-tolerance by preferentially supporting enhanced Treg function [71, 72], likely via an indirect mechanism affecting APCs [73, 74]. eHSP70 promotes functional Treg and their increased secretion of immunosuppressive TGF-β and IL-10 that may be accentuated in the presence of IL-2 [75]. Treg are similarly induced by vaccines integrating high doses of gp96 in mice [76]. High-dose tumor-derived gp96 binds to $CD91^{+}$ pDC, leading to (1) NFkB activation and translocation to the nucleus, (2) subsequent methylome remodeling and upregulation of NRP1 in pDC, and (3) promotion/activation of immunosuppressive Treg cells [77, 78]. Herber et al. have also reported that DC ($CD11c^{+}CD8a^{+}$ and conventional DC but not pDC) in tumor-bearing mice and humans exhibit dysfunction and high levels of triglycerides as a consequence of the selective DC upregulation of the SR-A (aka CD204) scavenger receptor [79], which is known to bind and internalize large HSPs (i.e., hsp110 and grp170), leading to suboptimal induction of antigen-specific T-cell responses [80].

This suggests that therapeutic strategies that are too efficient in promoting acute and extensive tumor cell apoptosis/necrosis could result in high levels of free eHSP release leading to the (counterintuitive and selective) promotion of regulatory DC function and Treg-dominated immune responses. Controlled therapy-induced tumor cell death (via the use of metronomic approaches) might instead be preferred, since more moderate amounts of purified tumor HSP70 and gp96 (which bind tumor-derived peptides) can been used in vaccine formulations to effectively (cross)prime therapeutic antitumor $CD8^{+}$ T-cell responses in vivo in both mice and cancer patients

[68]. In such cases, eHSPs can interact with $CD91^+$ DC, resulting in their maturation (in association with enhanced MHC presentation and secretion of pro-inflammatory cytokines/chemokines such as IL-12p70) and their ability to cross prime protective Type-1 $CD8^+$ T cells [81]. Such considerations might allow for the improvement of clinical trials implementing vaccines based on tumor-associated eHSP70 (mostly phase I) or eHSP90 (phase I–III) that have demonstrated immunogenicity but rarely lead to objective clinical responses [82–84].

## 9.4 Effects of HSP Inhibitors (HSPi) on Tumor Cell Immunogenicity In Vitro and In Vivo

A broad range of HSP inhibitors have been developed over the past several decades primarily in order to antagonize (at low nanomolar concentrations) the ability of HSP70/90 to sustain tumor cell expression of oncoproteins and "hallmark" pathways associated with tumor growth, survival, metastasis, and the genotoxic action of chemo- or radiotherapy, with 17 different HSP90i advancing to clinical trials for the treatment of more than 1000 cancer patients to date [85–90]. However, it has become increasingly clear that the functional antagonism of HSP(90) function on a systemic level will also impact the nature of tumor cell surveillance by the immune system, particularly in the therapeutically relevant TME [91–94].

The ability of tumor cells to be recognized by specific $CD4^+$ and $CD8^+$ T cells is dependent on tumor cell presentation of cognate peptides in the context of MHC II and MHC I molecules, respectively, with further "fine tuning" of T-cell-tumor interactions modulated by the action of tumor cell-expressed co-stimulatory/co-inhibitory molecules and adhesion molecules [95–98]. Since many tumor-associated antigens are overexpressed, non-mutated proteins that are also expressed by normal host somatic tissues, "self-tolerance" mechanisms typically restrict the tumor antigen-specific T-cell repertoire to clonotypes harboring only modest functional avidity for tumor cell recognition [99]. Notably, HSP90i have been reported to enhance tumor cell MHC molecule expression and/or tumor antigen-derived peptide presentation by MHC I/II molecules on the surface of tumor cells, thereby facilitating the ability of moderate avidity T cells to recognize and react against treated tumor cells vs. untreated tumor cells or normal cells [100–102]. Such conditional HSP90i-dependent alterations in the MHC-presented "peptidome" allow at least certain species to exceed the operational tolerance threshold of the host's T-cell repertoire, allowing for the activation and mobilization of a previously silent T effector cell cohort with potential to mediate antitumor activity. Modulations in tumor immunogenicity induced by interventional drugs such as HSP90i become even more important clinically given pervasive reports for MHC and antigen-processing/presentation abnormalities in heterogeneous cancer cell populations in vivo that correlate with low levels of MHC-peptide complexes on the tumor cell surface and with disease progression and poor patient survival [103–105].

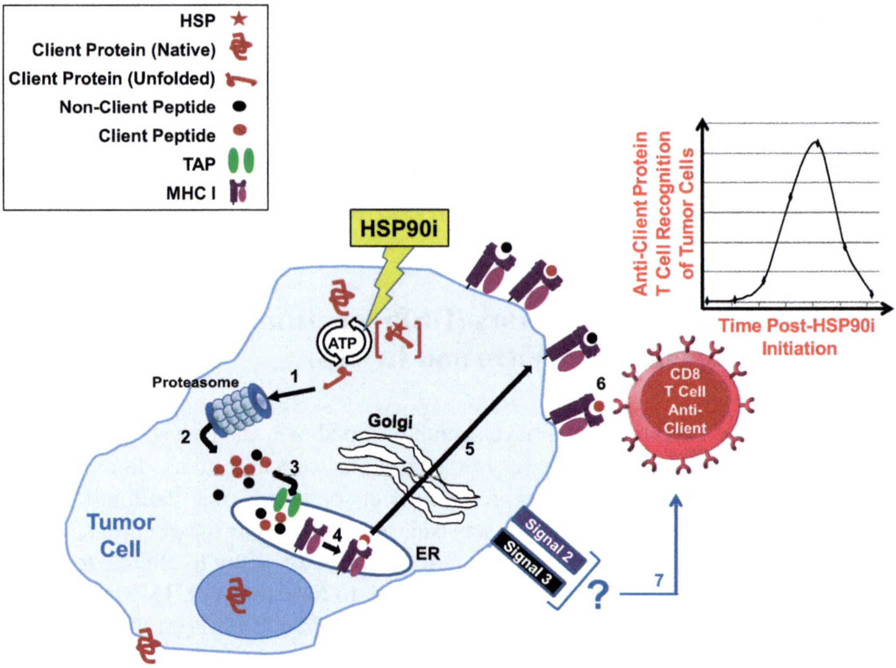


**Fig. 9.3** HSP90i promote the conditional processing and MHC class I presentation of client protein-derived T-cell epitopes in tumor cells. HSP90i interfering with ATP binding result in the delivery of client proteins to the multi-catalytic 26S proteasome complex (1) for degradation into peptides in the tumor cell cytoplasm (2). A fraction of this "wave" of client protein-derived peptides may be translocated into the tumor cell ER in a TAP-dependent manner (3) and loaded into nascent MHC class I complexes (4). Such MHC-client peptide complexes may then be shipped to the tumor cell surface (5), where they may be recognized by specific CD8+ T cells (6). Depending on additional signals conveyed by DC-to-T cells (i.e., signal 2 is contributed by the balance of co-stimulatory [CD40, CD70, CD80, CD86, 4-1BBL, OX40L, among others] versus co-inhibitory molecule interactions [PD-L1, PD-L2, among others], signal 3 reflects the balance between secreted pro-inflammatory versus regulatory/suppressive factors), the functional polarity of resultant CD8+ T effector cells may be imprinted. Notably, the ability of client peptide-specific CD8+ T cells to recognize HSP90i-treated tumor cells exhibits a temporal "prozone," with optimal recognition occurring several days after drug treatment [92, 100, 101]. Sustained delivery of HSP90i leads to subsequent diminishment in tumor cell recognition by CTL. The reasons for specific T-cell hyporeactivity at later time points are likely to be multifactorial and may reflect the confounding influence of compensatory HSP70 upregulation in tumor cells that limits client protein processing, the accumulated inhibitory action of HSP90i on proteasome function, and/or to drug-induced upregulation of immune checkpoint molecules, among other possibilities

One manifestation of this therapeutic strategy involves treatment with HSP90i to promote the conditional processing and MHC presentation of HSP90 client proteins in tumor cells (Fig. 9.3). In particular, many of these overexpressed/accumulated client proteins are critically involved in supporting cancer hallmark pathways (Fig. 9.1), making them less dispensable to tumor cells attempting to become antigen-loss variants in the face of specific immune selective pressure. One of the first validations of this concept was provided in a report by Ioannides and colleagues [106] in which a 20-h treatment of SKOV3.A2 human ovarian carcinoma cells with the HSP90i geldanamycin (GA) resulted in the enhanced proteasome-dependent processing and tumor cell HLA-A2 class I presentation of a peptide epitope (E75; derived from the oncogenic HSP90 client protein HER2) to specific cytotoxic $CD8^+$ T cells in vitro. Analogous results have since been reported for HSP90i-heightened $CD8^+$ T-cell targeting of the tumor cell overexpressed pro-metastatic HSP90 client proteins EphA2 and MET (Fig. 9.1) in vitro and in vivo. Kawabe et al. demonstrated that the treatment of human HLA-$A2^+$, $EphA2^+$ melanoma cells with HSP90i 17-DMAG enhanced their recognition by EphA2-specific $CD8^+$ T-cell lines and clones in vitro via a mechanism that was dependent on both the proteasome and TAP peptide translocation into the ER after the initial Sec61-dependent retro-translocation of EphA2 protein into the tumor cytoplasm [101]. Subsequent studies in mice using an MCA205 sarcoma model demonstrated that oral administration of 17-DMAG acted synergistically with specific vaccination against EphA2 or with adoptive anti-EphA2 $CD8^+$ T-cell therapy in slowing tumor growth, via a mechanism involving transiently enhanced T-cell recognition of not only $EphA2^+$ tumor cells but also $EphA2^+$ tumor-associated vascular endothelial cells (VEC), but not $EphA2^+$ VEC from the kidneys of these tumor-bearing animals [92–94]. Since recent findings suggest that VEC in the stressful milieu of the tumor microenvironment also overexpress HSPs (including HSP90) versus VEC in non-diseased organs, this may broaden the relevance of this therapeutic paradigm to include HSP90i-improved immune targeting of tumor stromal cell populations. Interestingly, the ability of HSP90i to serve as a beneficial co-therapy in these models was shown to be highly dose- and schedule-dependent, with an intermediate dose (15 mg/kg/day for 5 days) proving to be optimally therapeutic. Higher or lower doses of 17-DMAG for longer or shorter periods of time yielded less therapy benefit. Similar results have also been published by others [102], where low (2.5 μg) but not high (10 μg) doses of HSP90i GA improved the antitumor efficacy of DNA-based vaccines against HSP90i client proteins HER2 or MET, with GA determined to improve tumor infiltration by $CD8^+$ T cells and NK cells and to sensitize tumors to specific T cell-mediated killing. Intriguingly, many HSPs (HSP27, HSP60, HSP70, HSP90) are "self-clients" that encode MHC-presented peptide epitopes recognized by specific $CD4^+$ and $CD8^+$ T effector cells and/or by $CD4^+$ Treg [107–111]. Notably, anti-HSP90 $CD8^+$ T cells were shown to mediate significant anti-myeloma efficacy when applied in ACT approaches in Hu-SCID tumor models [109]. Moving forward, it will be most interesting to discern whether HSPi promotes improved recognition of tumor cells by HSP-specific T cells in vitro and in vivo.

It is important to note that HSP90i have also been reported to improve $CD8^+$ T-cell recognition of tumor cell antigens that are not direct HSP90 client proteins. Haggerty et al. provide a comprehensive report that 12 distinct HSP90i that antagonize HSP90 via different mechanisms are all capable of leading to the increased transcription and synthesis of melanocyte-lineage antigens (i.e., MART-1, gp100, and TRP-2) in murine B16 melanoma cells and two human glioma cell lines. HSP90i improved B16 recognition by anti-MART-1 T effector cells in vitro, with an optimal dose and schedule of 0.1–1 μg/ml for 3 days [100]. Although the authors also suggest that HSP90i treatment of tumor cells resulted in their increased expression of cell surface MHC class I molecules [100], other studies failed to discern similar HSP90i induction of tumor cell MHC expression [92–94, 101, 112, 113]. Since the impact of HSP90i is presumed to be indirect, it is possible that the kinetics of tumor cell alterations in MHC class I expression may vary between the various model systems, making this an important biomarker to consider in prospective translational and clinical studies implementing HSP90i.

Published studies have yet to describe an impact of HSPi on tumor cell expression of MHC class II molecules and their ability to present tumor antigen-derived peptides to cognate $CD4^+$ T cells, despite the knowledge that HSP90 plays an important role in antigen presentation to T lymphocytes via major histocompatibility complex class II molecules [114, 115]. It is also somewhat surprising that HSPi effects on tumor cell expression of co-stimulatory (i.e., CD40, CD70, CD80, CD86, GITR-L, OX40L, and 4-1BBL, among others; [116]) or co-inhibitory/immune checkpoint molecules (i.e., PD-L1, Galectin-9/LGALS9, HVEM, TIGIT ligands, VISTA, among others; [117]) have yet to be comprehensively investigated. These will clearly represent high-priority targets for future studies that could shape the design of combination clinical protocols involving the administration of HSPi and immune potentiating agents, such as anti-PD-L1.

## 9.5 HSPis Can Have a Negative Impact on T Cell-, NK Cell-, and DC-Mediated Immunity When Applied in High Doses (Fig. 9.4)

Although a clear clinical intent for applying HSPi is to interrupt the intrinsic array of pro-oncogenic signaling taking place within tumor cells themselves and within the stroma of the TME, immune cells also require HSPs to sustain intrinsic client proteins crucial to their expansion, differentiation, survival, and functionality. In particular, antigen-presenting cells, B/T effector cells, and NK cells can be negatively impacted in vitro and in vivo by the presence of (high-dose) HSPi, prompting some investigators to implement HSPi-based treatment strategies in the setting of autoimmunity and solid organ transplantation where inflammatory immunity is contraindicated and Treg responses are instead preferred [118–122].

Hence, HSP90i have been demonstrated to promote apoptosis or hyporesponsiveness (reduced cytotoxicity and/or production of effector cytokines/chemokines) in B cells, T cells, and NK cells in a dose-dependent manner [123–127]. At the molecular level, treatment of T cells with HSP90i promotes the degradation of a broad spectrum of intrinsic client proteins that includes the TCR, as well as TCR co-receptors (i.e., CD3, CD4, CD8), co-stimulatory molecules (i.e., CD28, CD40L), and TCR-proximal signaling (i.e., LAT, LCK, ZAP70) molecules [123, 128–130]. Treatment of T cells with HSP90i also results in reduced Th1 cell expression of

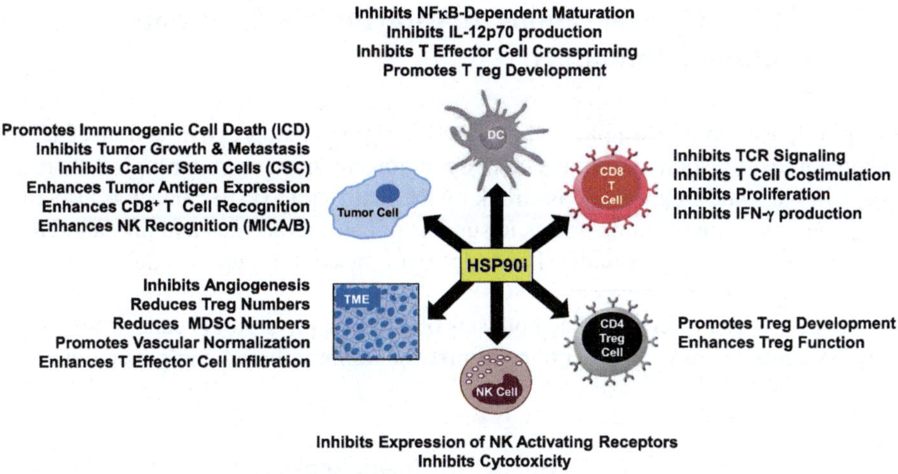


**Fig. 9.4** Dichotomous impact of HSP90i on cells within the TME. HSP90i have clearly demonstrated antitumor benefits based on their direct effects on tumor cells (i.e., inhibition of proliferation, promotion of apoptosis, suppression of cancer stem cells, enhancement of recognition by CD8+ T cells and NK cells) and tumor-associated stromal cells (i.e., inhibition of angiogenesis and the promotion of "vascular normalization," alteration in chemokine profiles resulting in enhanced Teff recruitment and reduced infiltration by suppressive Treg cells and MDSC). High, continuous dosing of HSP90i is contraindicated for optimal antitumor immune function, based on this regimen's known ability to inhibit proximal TCR signaling (based on its impact on client proteins such as Lck among others), leading to reduced T-cell proliferation, inhibition of co-stimulatory signaling and reduced capacity of Teff to produce inflammatory mediators such as IFN-γ, as well as to its predilection to support Treg-mediated immunosuppression (either directly or indirectly via regulatory effects on antigen-presenting cells). Sustained, high-dose application of HSP90i also perturbs innate cell effector functions in DC and NK cells. HSP90i can inhibit full DC maturation by interfering with TLR-mediated signaling and NFκB activation, resulting in DC that are deficient in their ability to (cross)prime Type-1 T-cell responses based on inhibition of DC IL-12p70 production and expression of co-stimulatory molecules and APM components. In NK cells, HSP90i have been reported to inhibit expression of NK-activating receptors (NKAR) and to limit the cytotoxic function of these effector cells

HSP90 client protein MTORC1 (Raptor) that inhibits memory responses, leading to T-cell dysfunction and/or anergy [131]. Innate immune function can also be impaired by HSP90i, since NK cell conditioning with GA leads to diminished expression of a range of cell surface-activating receptors, including CD2, CD11a, CD94, NKp30, NKp44, NKp46, and KARp50.3 [123]. Furthermore, eHSP70-facilitated cross presentation of the tumor oncofetal antigen 5T4 by CD91+ DC can be prevented by HSP70i [132], and others have reported that HSP90i preclude optimal DC maturation by limiting NFκB activation [126, 127], resulting in only feeble stimulation of cognate T-cell responses [125]. DC expression of MHC class I molecules can also be reduced in the presence of high-dose (1 μM for 24–48 h) HSP90i GA in vitro [133].

## 9.6 Clinical Considerations in Applying HSPis to Enhance Antitumor Immunity

Despite these many examples supporting the immunosuppressive action of HSPi when applied to isolated immune cell populations in vitro, or under high-dose regimens in vitro and in vivo, one is struck by the reality that HSPi can act as effective "adjuvants" or conditioning agents in support of cancer vaccines and immunotherapies when using metronomic (lower-dose) or discontinuous schedule regimens (Fig. 9.5; [92–94, 102]).

So, how does one consolidate both sets of data in a physiologic cancer immunology "systems biology" in order to maximize the immune-based component inherent

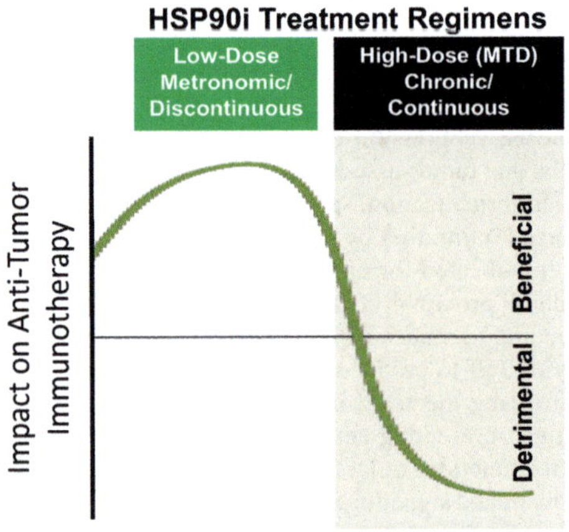


**Fig. 9.5** Recommended use of HSP90i in low-dose metronomic/discontinuous regimens for optimal immunotherapeutic benefit in the cancer setting. Given the myriad of client proteins impacted by HSP90i in tumor cells, tumor-associated stromal cells, and protective immune cells within the TME (and peripheral tissue compartments of a treated patient), HSP90i treatment regimens associated with optimal, durable clinical benefits will likely involve dosing far below established MTDs for this drug class, with only intermittent periods of drug delivery. Such dosing/scheduling will allow for a clinically preferred balance to be achieved between drug-associated suppression of the tumor-intrinsic "hallmarks of cancer" while limiting collateral inhibition of immune cell function that appears increasingly relevant to the mechanism of action underlying the antitumor efficacy of many chemotherapeutic agents. In particular, lower-dose, intermittent delivery of HSP90i may promote vascular normalization and the recruitment of anti-client protein CTL into the TME where they may better recognize cognate MHC class I-presented peptide complexes on the tumor cell surface, in a manner unopposed by regulatory cell populations such as Treg and MDSC. Vascular normalization may also result in improved DC infiltration of the TME, allowing for improved cross priming of an expanded antitumor T-cell repertoire over time (i.e., epitope spreading), which has been associated with durable objective clinical responses in patients treated using a range of immunotherapy approaches. Ideally, such low-dose HSP90i regimens might be combined with immunotherapies employing immune checkpoint blockade antibodies and/or ACT using patient T cells expanded (or engineered to express TCR reactive) against tumor overexpressed HSP90 client proteins, such as EphA2 or Her2/neu, among many others

to the antitumor efficacy associated with HSPi-based therapies? Firstly, HSPis tend to be sequestered/retained in the TME vs. plasma for extended periods of time after systemic delivery into patients or tumor-bearing mice [134–138]. This suggests that drug doses (far below the MTD) provided on an intermittent schedule would likely be required to achieve a sustained level of HSPi that optimizes the antitumor efficacy of immune cells within the TME. Although such treatment regimens might not optimize the ability of HSPi to inhibit tumor-intrinsic hallmarks of cancer (Fig. 9.1) in vivo, the ability to provide synergistic coverage in antagonizing "more" hallmarks may lead to greater durability in HSPi therapy benefits due to memory developed in the adaptive antitumor immune response. Less acute, catastrophic killing of tumor cells would theoretically limit the quantity of free eHSPs released, avoiding HSP-R (i.e., CD91)-mediated regulatory signals in DC [77, 78] in support of improved antitumor T effector cell cross priming/restimulation.

Beyond considering the simple interplay of tumor cells and their products (HSP-related or otherwise) with isolated immune cells, one must consider the broader implications of HSPi impact on the field of battle in which the protective immune cells combat cancer cells, i.e., the TME. Notably, the progressor TME is both (1) intrinsically poor in recruiting protective immune cell populations and (2) hostile to the survival/function of infiltrating immune (nonregulatory) effector cells based on the suppressive impact of hypoxia, acidosis, high interstitial fluid pressure, MDSC, and Treg [139–142]. All of these suppressive indices can be made more favorable as a consequence of "vascular normalization" (VN) driven by antiangiogenic agents targeting EGFR-, HER2-, VEGFR-, and/or PDGFR-mediated signaling pathways in tumor-associated blood vessel cells and their recruited precursor cells [143–148]. The process of VN involves the trimming of chaotic vascular arborization in the TME, improved pericyte coverage of residual vascular conduits, reestablishment of tissue normoxia, and improved vascular integrity and tissue perfusion (with reduced acidosis and enhanced infiltration by immune effector cells in response to inflammatory chemokines produced in the tumor stroma). From an immunologic perspective, VN is also associated with reductions in the levels of MDSC and Treg within the TME [149]. Since the major VN targets represent known HSP90 client proteins (Fig. 9.1), it should not come as a surprise that treatment with (low-dose) HSP90i has been reported to optimally promote VN in the tumor lesions of treated individuals (Fig. 9.4; [92, 93]). In particular, the treatment of established murine tumors with low doses of orally administered HSP90i (17-DMAG) for no more than 5 days resulted in coordinate immunotherapeutic benefits including (1) slowed tumor growth or regression in association with reduced vascular arborization, (2) activation of stromal cell production of CXCR3 ligand chemokines known to recruit Type-1 T effector cells, (3) activation of vascular endothelial cells to express VCAM1 used to facilitate Type-1 T-cell extravasation into the TME, (4) improved levels of antitumor $CD8^+$ TIL, (5) reduction in levels of MDSC and Treg in the TME, and (6) enhanced/prolonged recognition of tumor cells by $CD8^+$ T cells reactive against the HSP90 client protein EphA2 [92].

## 9.7 Conclusions and Future Directions

HSPs interact with, and stabilize, an ever-increasing number of client proteins that support normal cellular homeostasis. However, in malignant cells HSPs are commonly overexpressed and subjugated to extend the fate of an array of pro-tumor and anti-immune client proteins that serve as markers of poor disease prognosis. Since HSPs are ubiquitously expressed, it comes as little surprise that HSPis, when applied at high doses, exert undesired "on-target" toxicities to normal cell populations, including immune cells (i.e., DC, T cells, and NK cells, among others). In the case of HSP90i, high-dose treatment regimens targeting both tumor cells and tumor-associated blood vessels may lead to catastrophic tumor cell death and the release of tumor eHSPs in large quantities, leading to DC subversion and the induction of regulatory (pro-tumor) immune responses, rather than the desired induction of protective T and NK cell responses. Since systemically administered HSPis are retained for extended periods of time in the TME, where the operation of therapeutic immune cells must be conserved for optimal clinical benefit, dosing/scheduling regimens must be carefully selected to avoid exceeding thresholds associated with protective DC, T-cell, and NK cell inactivation.

The ability of HSPi to promote VN also implies that systemically applied drug will be delivered into the TME more efficiently [149], which might suggest the consideration of protocols involving HSPi dose de-escalation over an extended period of treatment in order to optimize the immune-stimulatory component associated with HSPi-based therapies. Since HSPis promote VN, leading to T effector cell recruitment, logical combination immunotherapy protocols could integrate active vaccination to elicit increased frequencies of circulating antitumor T effector cells (reactive against tumor-associated proteins including HSP client proteins). To minimize any deleterious effects of HSPi on vaccine priming of antitumor T cells in vivo, one could also implement HSPi combination therapies involving the use of adoptively transferred antitumor (ex vivo expanded or genetically engineered to express recombinant TCRs or chimeric antigen receptor (CAR)s). Furthermore, the antitumor efficacy of these approaches would be expected to be bolstered by the additional inclusion of immune checkpoint blockade (ICB) via the administration of antagonist antibodies against PD-1, PD-L1, and CTLA4, among others [150]. Given the profound autoimmune sequela associated with ICB-based therapies, such toxicities will need to be carefully monitored in any prospective combination HSPi + ICB protocols.

In closing, it is also tempting to note that strikingly similar immunomodulatory effects have been observed for histone deacetylase inhibitors (HDACi) that can also operationally serve as HSP90i given their ability to promote the hyperacetylation of lysine residues in HSPs that are critical to its function as a chaperone molecule [151–153]. Like HSPi, HDACi have been reported to exert both immunostimulatory and immunosuppressive effects in vitro and in vivo [154], with recent reports suggesting that HDACi coordinately enhance DC, NK, and T-cell function against tumor cells [155, 156]. HDACi can increase tumor cell sensitivity to attack mediated by T and NK cells based on tumor cell upregulation of MHC/CS or NK/NKT cell ligands ([94, 157–159]) and protect T cells in the TME from apoptotic cell death by suppressing Fas expression on T effector cells [160]. Furthermore, HDACi have been found to synergize with immunotherapies (i.e., vaccines and ACT approaches) to yield objective responses in the cancer setting [158, 161, 162]. Given their dichotomous effects on the immune system, the clinical use of HDACi to augment therapeutic antitumor immunity would likely be subject to the same considerations for reduced dose and intermittent scheduling as has been previously suggested for HSPi applications. Furthermore, since HDACi have been reported to upregulate checkpoint molecules (such as PD-L1 on tumor cells; [163]), it may be particularly cogent to consider co-therapies using ICB in the clinical setting at outset.

**Acknowledgments** This work was supported by NIH R01 CA169118 (WJS) and NIH P50 CA121973 CA (Career Enhancement Award to RJF).

## References

1. Calderwood SK (2012) HSF1, a versatile factor in tumorogenesis. Curr Mol Med 12:1102–1107
2. Ciocca DR, Calderwood SK (2005) Heat shock proteins in cancer: diagnostic, prognostic, predictive, and treatment implications. Cell Stress Chaperones 10:86–103
3. Franceschelli S, Rosati A, Lerose R, De Nicola S, Turco MC, Pascale M (2008) Bag3 gene expression is regulated by heat shock factor 1. J Cell Physiol 215:575–577. https://doi.org/10.1002/jcp.21397
4. Garrido C, Brunet M, Didelot C, Zermati Y, Schmitt E, Kroemer G (2006) Heat shock proteins 27 and 70: anti-apoptotic proteins with tumorigenic properties. Cell Cycle 5:2592–2601
5. Ischia J, So AI (2013) The role of heat shock proteins in bladder cancer. Nat Rev Urol 10:386–395. https://doi.org/10.1038/nrurol.2013.108
6. Jäättelä M (1999) Heat shock proteins as cellular lifeguards. Ann Med 31:261–271
7. Jiang S, Tu K, Fu Q, Schmitt DC, Zhou L, Lu N et al (2015a) Multifaceted roles of HSF1 in cancer. Tumour Biol 36:4923–4931. https://doi.org/10.1007/s13277-015-3674-x
8. Kamal A, Thao L, Sensintaffar J, Zhang L, Boehm MF, Fritz LC et al (2003) A high-affinity conformation of Hsp90 confers tumour selectivity on Hsp90 inhibitors. Nature 425:407–410
9. Kimura A, Ogata K, Altan B, Yokobori T, Ide M, Mochiki E et al (2016) Nuclear heat shock protein 110 expression is associated with poor prognosis and chemotherapy resistance in gastric cancer. Oncotarget 7:18415–18423. https://doi.org/10.18632/oncotarget.7821
10. Prodromou C (2016) Mechanisms of Hsp90 regulation. Biochem J 473:2439–2452. https://doi.org/10.1042/BCJ20160005
11. Rérole AL, Jego G, Garrido C (2011) Hsp70: anti-apoptotic and tumorigenic protein. Methods Mol Biol 787:205–230. https://doi.org/10.1007/978-1-61779-295-3_16
12. Tang D, Khaleque MA, Jones EL, Theriault JR, Li C, Wong WH et al (2005) Expression of heat shock proteins and heat shock protein messenger ribonucleic acid in human prostate carcinoma in vitro and in tumors in vivo. Cell Stress Chaperones 10:46–58
13. Vydra N, Toma A, Glowala-Kosinska M, Gogler-Piglowska A, Widlak W (2013) Overexpression of heat shock transcription factor 1 enhances the resistance of melanoma cells to doxorubicin and paclitaxel. BMC Cancer 13:504. https://doi.org/10.1186/1471-2407-13-504

14. Jiang L, Kwong DL, Li Y, Liu M, Yuan YF, Li Y et al (2015b) HBP21, a chaperone of heat shock protein 70, functions as a tumor suppressor in hepatocellular carcinoma. Carcinogenesis 36:1111–1120. https://doi.org/10.1093/carcin/bgv116
15. Zhang TT, Jiang YY, Shang L, Shi ZZ, Liang JW, Wang Z et al (2015) Overexpression of DNAJB6 promotes colorectal cancer cell invasion through an IQGAP1/ERK-dependent signaling pathway. Mol Carcinog 54:1205–1213. https://doi.org/10.1002/mc.22194
16. Cornford PA, Dodson AR, Parsons KF, Desmond AD, Woolfenden A, Fordham M et al (2000) Heat shock protein expression independently predicts clinical outcome in prostate cancer. Cancer Res 60:7099–7105
17. Foster CS, Dodson AR, Ambroisine L, Fisher G, Møller H, Clark J et al (2009) Hsp-27 expression at diagnosis predicts poor clinical outcome in prostate cancer independent of ETS-gene rearrangement. Br J Cancer 101:1137–1144. https://doi.org/10.1038/sj.bjc.6605227
18. Li S, Zhang W, Fan J, Lai Y, Che G (2016) Clinicopathological and prognostic significance of heat shock protein 27 (HSP27) expression in non-small cell lung cancer: a systematic review and meta-analysis. Springerplus 5:1165. https://doi.org/10.1186/s40064-016-2827-8
19. Thomas X, Campos L, Mounier C, Cornillon J, Flandrin P, Le QH et al (2005) Expression of heat-shock proteins is associated with major adverse prognostic factors in acute myeloid leukemia. Leuk Res 29:1049–1058
20. Yang Z, Zhuang L, Szatmary P, Wen L, Sun H, Lu Y et al (2015a) Upregulation of heat shock proteins (HSPA12A, HSP90B1, HSPA4, HSPA5 and HSPA6) in tumour tissues is associated with poor outcomes from HBV-related early-stage hepatocellular carcinoma. Int J Med Sci 12:256–263. https://doi.org/10.7150/ijms.10735
21. Cheng Q, Chang JT, Geradts J, Neckers LM, Haystead T, Spector NL et al (2012) Amplification and high-level expression of heat shock protein 90 marks aggressive phenotypes of human epidermal growth factor receptor 2 negative breast cancer. Breast Cancer Res 14:R62
22. Kang GH, Lee EJ, Jang KT, Kim KM, Park CK, Lee CS et al (2010) Expression of HSP90 in gastrointestinal stromal tumours and mesenchymal tumours. Histopathology 56:694–701. https://doi.org/10.1111/j.1365-2559.2010.03550.x
23. Li XS, Xu Q, Fu XY, Luo WS (2014a) Heat shock protein 60 overexpression is associated with the progression and prognosis in gastric cancer. PLoS One 9:e107507. https://doi.org/10.1371/journal.pone.0107507
24. Collura A, Lagrange A, Svrcek M, Marisa L, Buhard O, Guilloux A et al (2014) Patients with colorectal tumors with microsatellite instability and large deletions in HSP110 T17 have improved response to 5-fluorouracil–based chemotherapy. Gastroenterology 146:401–11.e1
25. Dai C, Sampson SB (2016) HSF1: guardian of proteostasis in cancer. Trends Cell Biol 26:17–28. https://doi.org/10.1016/j.tcb.2015.10.011
26. Li S, Ma W, Fei T, Lou Q, Zhang Y, Cui X et al (2014b) Upregulation of heat shock factor 1 transcription activity is associated with hepatocellular carcinoma progression. Mol Med Rep 10:2313–2321. https://doi.org/10.3892/mmr.2014.2547
27. Taira T, Negishi Y, Kihara F, Iguchi-Ariga SM, Ariga H (1992) c-myc protein complex binds to two sites in human hsp70 promoter region. Biochim Biophys Acta 1130:166–174
28. Calderwood SK, Xie Y, Wang X, Khaleque MA, Chou SD, Murshid A et al (2010) Signal transduction pathways leading to heat shock transcription. Sign Transduct Insights 2:13–24
29. Home T, Jensen RA, Rao R (2015) Heat shock factor 1 in protein homeostasis and oncogenic signal integration. Cancer Res 75:907–912. https://doi.org/10.1158/0008-5472.CAN-14-2905
30. Heimberger T, Andrulis M, Riedel S, Stühmer T, Schraud H, Beilhack A et al (2013) The heat shock transcription factor 1 as a potential new therapeutic target in multiple myeloma. Br J Haematol 160:465–476. https://doi.org/10.1111/bjh.12164
31. Taipale M, Jarosz DF, Lindquist S (2010) HSP90 at the hub of protein homeostasis: emerging mechanistic insights. Nat Rev Mol Cell Biol 11:515–528. https://doi.org/10.1038/nrm2918
32. Hanahan D, Weinberg RA (2011) Hallmarks of cancer: the next generation. Cell 144:646–674. https://doi.org/10.1016/j.cell.2011.02.013

33. Miyata Y, Nakamoto H, Neckers L (2013) The therapeutic target Hsp90 and cancer hallmarks. Curr Pharm Des 19:347–365
34. Neckers L, Workman P (2012) Hsp90 molecular chaperone inhibitors: are we there yet? Clin Cancer Res 18:64–76. https://doi.org/10.1158/1078-0432.CCR-11-1000
35. Hanahan D, Weinberg RA (2000) Cell 100:57–70
36. Wang SC, Lien HC, Xia W, Chen IF, Lo HW, Wang Z et al (2004) Binding at and transactivation of the COX-2 promoter by nuclear tyrosine kinase receptor ErbB-2. Cancer Cell 6:251–261
37. Mohammadi A, Yaghoobi MM, Gholamhoseinian Najar A, Kalantari-Khandani B, Sharifi H, Saravani M (2016) HSP90 inhibition suppresses PGE2 production via modulating COX-2 and 15-PGDH expression in HT-29 colorectal cancer cells. Inflammation 39:1116–1123. https://doi.org/10.1007/s10753-016-0343-1
38. Tanioka T, Nakatani Y, Kobayashi T, Tsujimoto M, Oh-ishi S, Murakami M et al (2003) Regulation of cytosolic prostaglandin E2 synthase by 90-kDa heat shock protein. Biochem Biophys Res Commun 303:1018–1023
39. Obermajer N, Wong JL, Edwards RP, Odunsi K, Moysich K, Kalinski P (2012) PGE(2)-driven induction and maintenance of cancer-associated myeloid-derived suppressor cells. Immunol Invest 41:635–657. https://doi.org/10.3109/08820139.2012.695417
40. Sahin M, Sahin E, Koksoy S (2013) Regulatory T cells in cancer: an overview and perspectives on cyclooxygenase-2 and Foxp3 DNA methylation. Hum Immunol 74:1061–1068. https://doi.org/10.1016/j.humimm.2013.05.009
41. Agrawal B, Krantz MJ, Reddish MA, Longenecker BM (1998) Cancer-associated MUC1 mucin inhibits human T-cell proliferation, which is reversible by IL-2. Nat Med 4:43–49
42. Chan AK, Lockhart DC, von Bernstorff W, Spanjaard RA, Joo HG, Eberlein TJ et al (1999) Soluble MUC1 secreted by human epithelial cancer cells mediates immune suppression by blocking T-cell activation. Int J Cancer 82:721–726
43. Wang J, He P, Gaida M, Yang S, Schetter AJ, Gaedcke J et al (2016) Inducible nitric oxide synthase enhances disease aggressiveness in pancreatic cancer. Oncotarget 7:52993. https://doi.org/10.18632/oncotarget.10323
44. Yang L, Wang Y, Guo L, Wang L, Chen W, Shi B (2015b) The expression and correlation of iNOS and p53 in oral squamous cell carcinoma. Biomed Res Int 2015:637853. https://doi.org/10.1155/2015/637853
45. Zhang W, He XJ, Ma YY, Wang HJ, Xia YJ, Zhao ZS et al (2011) Inducible nitric oxide synthase expression correlates with angiogenesis, lymphangiogenesis, and poor prognosis in gastric cancer patients. Hum Pathol 42:1275–1282. https://doi.org/10.1016/j.humpath.2010.09.020
46. Jayaraman P, Parikh F, Lopez-Rivera E, Hailemichael Y, Clark A, Ma G et al (2012) Tumor-expressed inducible nitric oxide synthase controls induction of functional myeloid-derived suppressor cells through modulation of vascular endothelial growth factor release. J Immunol 188:5365–5376. https://doi.org/10.4049/jimmunol.1103553
47. Singer K, Gottfried E, Kreutz M, Mackensen A (2011) Suppression of T-cell responses by tumor metabolites. Cancer Immunol Immunother 60:425–431. https://doi.org/10.1007/s00262-010-0967-1
48. Sun X, Sui Q, Zhang C, Tian Z, Zhang J (2013) Targeting blockage of STAT3 in hepatocellular carcinoma cells augments NK cell functions via reverse hepatocellular carcinoma-induced immune suppression. Mol Cancer Ther 12:2885–2896. https://doi.org/10.1158/1535-7163.MCT-12-1087
49. Zhou YJ, Binder RJ (2014) The heat shock protein-CD91 pathway mediates tumor immune-surveillance. Oncoimmunology 3:e28222
50. Calderwood SK, Gong J, Murshid A (2016) Extracellular HSPs: the complicated roles of extracellular HSPs in immunity. Front Immunol 7:159. https://doi.org/10.3389/fimmu.2016.00159
51. Tamura Y, Torigoe T, Kukita K, Saito K, Okuya K, Kutomi G et al (2012) Heat-shock proteins as endogenous ligands building a bridge between innate and adaptive immunity. Immunotherapy 4:841–852. https://doi.org/10.2217/imt.12.75

52. Basu S, Binder RJ, Ramalingam T, Srivastava PK (2001) CD91 is a common receptor for heat shock proteins gp96, hsp90, hsp70, and calreticulin. Immunity 14:303–313
53. Dai J, Liu B, Caudill MM, Zheng H, Qiao Y, Podack ER et al (2003) Cell surface expression of heat shock protein gp96 enhances cross-presentation of cellular antigens and the generation of tumor-specific T cell memory. Cancer Immun 3:1
54. Kepp O, Senovilla L, Vitale I, Vacchelli E, Adjemian S, Agostinis P et al (2014) Consensus guidelines for the detection of immunogenic cell death. Oncoimmunology 3:e955691
55. Kroemer G, Galluzzi L, Kepp O, Zitvogel L (2013) Immunogenic cell death in cancer therapy. Annu Rev Immunol 31:51–72. https://doi.org/10.1146/annurev-immunol-032712-100008
56. Zhu H, Fang X, Zhang D, Wu W, Shao M, Wang L et al (2016) Membrane-bound heat shock proteins facilitate the uptake of dying cells and cross-presentation of cellular antigen. Apoptosis 21:96–109. https://doi.org/10.1007/s10495-015-1187-0
57. Chang CL, Hsu YT, CC W, Yang YC, Wang C, TC W et al (2012) Immune mechanism of the antitumor effects generated by bortezomib. J Immunol 189:3209–3220. https://doi.org/10.4049/jimmunol.1103826
58. Fucikova J, Kralikova P, Fialova A, Brtnicky T, Rob L, Bartunkova J et al (2011) Human tumor cells killed by anthracyclines induce a tumor-specific immune response. Cancer Res 71:4821–4833. https://doi.org/10.1158/0008-5472.CAN-11-0950
59. Multhoff G, Pockley AG, Schmid TE, Schilling D (2015) The role of heat shock protein 70 (Hsp70) in radiation-induced immunomodulation. Cancer Lett 368:179–184. https://doi.org/10.1016/j.canlet.2015.02.013
60. Panzarini E, Inguscio V, Dini L (2013) Immunogenic cell death: can it be exploited in PhotoDynamic Therapy for cancer? Biomed Res Int 2013:482160. https://doi.org/10.1155/2013/482160.
61. Multhoff G, Botzler C, Jennen L, Schmidt J, Ellwart J, Issels R (1997) Heat shock protein 72 on tumor cells: a recognition structure for natural killer cells. J Immunol 158:4341–4350
62. Chalmin F, Ladoire S, Mignot G, Vincent J, Bruchard M, Remy-Martin JP et al (2010) Membrane-associated Hsp72 from tumor-derived exosomes mediates STAT3-dependent immunosuppressive function of mouse and human myeloid-derived suppressor cells. J Clin Invest 120:457–471. https://doi.org/10.1172/JCI40483
63. Diao J, Yang X, Song X, Chen S, He Y, Wang Q et al (2015) Exosomal Hsp70 mediates immunosuppressive activity of the myeloid-derived suppressor cells via phosphorylation of Stat3. Med Oncol 32:453. https://doi.org/10.1007/s12032-014-0453-2
64. Gobbo J, Marcion G, Cordonnier M, Dias AM, Pernet N, Hammann A et al (2015) Restoring anticancer immune response by targeting tumor-derived exosomes with a HSP70 peptide aptamer. J Natl Cancer Inst 108(3). https://doi.org/10.1093/jnci/djv330. pii: djv330
65. Cho JA, Lee YS, Kim SH, Ko JK, Kim CW (2009) MHC independent anti-tumor immune responses induced by Hsp70-enriched exosomes generate tumor regression in murine models. Cancer Lett 275:256–265. https://doi.org/10.1016/j.canlet.2008.10.021
66. Binder RJ, Zhou YJ, Messmer MN, Pawaria S (2012) CD91-dependent modulation of immune responses by heat shock proteins: a role in autoimmunity. Autoimmune Dis 2012:863041. https://doi.org/10.1155/2012/863041
67. Chandawarkar RY, Wagh MS, Kovalchin JT, Srivastava P (2004) Immune modulation with high-dose heat-shock protein gp96: therapy of murine autoimmune diabetes and encephalomyelitis. Int Immunol 16:615–624
68. Chandawarkar RY, Wagh MS, Srivastava PK (1999) The dual nature of specific immunological activity of tumor-derived gp96 preparations. J Exp Med 189:1437–1442
69. Akyol S, Gercel-Taylor C, Reynolds LC, Taylor DD (2006) HSP-10 in ovarian cancer: expression and suppression of T-cell signaling. Gynecol Oncol 101:481–486
70. Banerjee S, Lin CF, Skinner KA, Schiffhauer LM, Peacock J, Hicks DG et al (2011) Heat shock protein 27 differentiates tolerogenic macrophages that may support human breast cancer progression. Cancer Res 71:318–327. https://doi.org/10.1158/0008-5472.CAN-10-1778

71. van Roon JA, van Eden W, van Roy JL, Lafeber FJ, Bijlsma JW (1997) Stimulation of suppressive T cell responses by human but not bacterial 60-kD heat-shock protein in synovial fluid of patients with rheumatoid arthritis. J Clin Invest 100:459–463
72. Zonneveld-Huijssoon E, Roord ST, de Jager W, Klein M, Albani S, Anderton SM et al (2011) Bystander suppression of experimental arthritis by nasal administration of a heat shock protein peptide. Ann Rheum Dis 70:2199–2206. https://doi.org/10.1136/ard.2010.136994
73. Kilmartin B, Reen DJ (2004) HSP60 induces self-tolerance to repeated HSP60 stimulation and cross-tolerance to other pro-inflammatory stimuli. Eur J Immunol 34:2041–2051
74. Zhong Y, Tang H, Wang X, Zeng Q, Liu Y, Zhao XI et al (2016) Intranasal immunization with heat shock protein 60 induces $CD4^+$ $CD25^+$ $GARP^+$ and type 1 regulatory T cells and inhibits early atherosclerosis. Clin Exp Immunol 183:452–468. https://doi.org/10.1111/cei.12726
75. Wachstein J, Tischer S, Figueiredo C, Limbourg A, Falk C, Immenschuh S et al (2012) HSP70 enhances immunosuppressive function of $CD4^+CD25^+FoxP3^+$ T regulatory cells and cytotoxicity in $CD4^+CD25^{neg}$ T cells. PLoS One 7:e51747. https://doi.org/10.1371/journal.pone.0051747
76. Li X, Liu Z, Yan X, Zhang X, Li Y, Zhao B et al (2013) Induction of regulatory T cells by high-dose gp96 suppresses murine liver immune hyperactivation. PLoS One 8:e68997. https://doi.org/10.1371/journal.pone.0068997
77. De Filippo A, Binder RJ, Camisaschi C, Beretta V, Arienti F, Villa A et al (2008) Human plasmacytoid dendritic cells interact with gp96 via CD91 and regulate inflammatory responses. J Immunol 181:6525–6535
78. Kinner L, Sedlacek AL, Watkins SC, Binder RJ (2016) Gp96 initiates DNA methylome remodeling in dendritic cells and facilitates interaction with T cells. J Immunol 196:S59.25
79. Herber DL, Cao W, Nefedova Y, Novitskiy SV, Nagaraj S, Tyurin VA et al (2010) Lipid accumulation and dendritic cell dysfunction in cancer. Nat Med 16:880–886. https://doi.org/10.1038/nm.2172
80. Qian J, Yi H, Guo C, Yu X, Zuo D, Chen X et al (2011) CD204 suppresses large heat shock protein-facilitated priming of tumor antigen gp100-specific T cells and chaperone vaccine activity against mouse melanoma. J Immunol 187:2905–2914. https://doi.org/10.4049/jimmunol.1100703
81. Shevtsov M, Multhoff G (2016) Heat shock protein-peptide and HSP-based immunotherapies for the treatment of cancer. Front Immunol 7:171. https://doi.org/10.3389/fimmu.2016.00171
82. Li JL, Liu HL, Zhang XR, JP X, WK H, Liang M et al (2009) A phase I trial of intratumoral administration of recombinant oncolytic adenovirus overexpressing HSP70 in advanced solid tumor patients. Gene Ther 16:376–382. https://doi.org/10.1038/gt.2008.179
83. Maeda Y, Yoshimura K, Matsui H, Shindo Y, Tamesa T, Tokumitsu Y et al (2015) Dendritic cells transfected with heat-shock protein 70 messenger RNA for patients with hepatitis C virus-related hepatocellular carcinoma: a phase 1 dose escalation clinical trial. Cancer Immunol Immunother 64:1047–1056. https://doi.org/10.1007/s00262-015-1709-1
84. Shevtsov MA, Kim AV, Samochernych KA, Romanova IV, Margulis BA, Guzhova IV et al (2014) Pilot study of intratumoral injection of recombinant heat shock protein 70 in the treatment of malignant brain tumors in children. Onco Targets Ther 7:1071–1081. https://doi.org/10.2147/OTT.S62764
85. He S, Smith DL, Sequeira M, Sang J, Bates RC, Proia DA (2014) The HSP90 inhibitor ganetespib has chemosensitizer and radiosensitizer activity in colorectal cancer. Invest New Drugs 32:577–586. https://doi.org/10.1007/s10637-014-0095-4
86. Isaacs JS (2016) Hsp90 as a "Chaperone" of the epigenome: insights and opportunities for cancer therapy. Adv Cancer Res 129:107–140. https://doi.org/10.1016/bs.acr.2015.09.003
87. Neckers L, Ivy SP (2003) Heat shock protein 90. Curr Opin Oncol 15:419–424
88. Proia DA, Kaufmann GF (2015) Targeting heat-shock protein 90 (HSP90) as a complementary strategy to immune checkpoint blockade for cancer therapy. Cancer Immunol Res 3:583–589. https://doi.org/10.1158/2326-6066.CIR-15-0057

89. Trepel J, Mollapour M, Giaccone G, Neckers L (2010) Targeting the dynamic HSP90 complex in cancer. Nat Rev Cancer 10:537–549. https://doi.org/10.1038/nrc2887
90. Wang Y, Trepel JB, Neckers LM, Giaccone G (2010) STA-9090, a small-molecule Hsp90 inhibitor for the potential treatment of cancer. Curr Opin Investig Drugs 11:1466–1476
91. Graner MW (2016) HSP90 and immune modulation in cancer. Adv Cancer Res 129:191–224. https://doi.org/10.1016/bs.acr.2015.10.001
92. Rao A, Taylor JL, Chi-Sabins N, Kawabe M, Gooding WE, Storkus WJ (2012a) Combination therapy with HSP90 inhibitor 17-DMAG reconditions the tumor microenvironment to improve recruitment of therapeutic T cells. Cancer Res 72:3196–3206. https://doi.org/10.1158/0008-5472.CAN-12-0538
93. Rao A, Lowe DB, Storkus WJ (2012b) Shock block for improved immunotherapy. Oncoimmunology 1:1427–1429
94. Rao R, Fiskus W, Ganguly S, Kambhampati S, Bhalla KN (2012c) HDAC inhibitors and chaperone function. Adv Cancer Res 116:239–262. https://doi.org/10.1016/B978-0-12-394387-3.00007-0
95. Garcia-Lora A, Algarra I, Garrido F (2003) MHC class I antigens, immune surveillance, and tumor immune escape. J Cell Physiol 195:346–355
96. Kessler BM, Glas R, Ploegh HL (2002) MHC class I antigen processing regulated by cytosolic proteolysis-short cuts that alter peptide generation. Mol Immunol 39:171–179
97. Storkus WJ, Herrem C, Kawabe M, Cohen PA, Bukowski RM, Finke JH et al (2007) Improving immunotherapy by conditionally enhancing MHC class I presentation of tumor antigen-derived Peptide epitopes. Crit Rev Immunol 27:485–493
98. Wang RF (2001) The role of MHC class II-restricted tumor antigens and CD4$^{+}$ T cells in antitumor immunity. Trends Immunol 22:269–276
99. Makkouk A, Weiner GJ (2015) Cancer immunotherapy and breaking immune tolerance: new approaches to an old challenge. Cancer Res 75:5–10. https://doi.org/10.1158/0008-5472.CAN-14-2538
100. Haggerty TJ, Dunn IS, Rose LB, Newton EE, Pandolfi F, Kurnick JT (2014) Heat shock protein-90 inhibitors enhance antigen expression on melanomas and increase T cell recognition of tumor cells. PLoS One 9:e114506. https://doi.org/10.1371/journal.pone.0114506
101. Kawabe M, Mandic M, Taylor JL, Vasquez CA, Wesa AK, Neckers LM et al (2009) Heat shock protein 90 inhibitor 17-dimethylaminoethylamino-17-demethoxygeldanamycin enhances EphA2$^{+}$ tumor cell recognition by specific CD8+ T cells. Cancer Res 69:6995–7003. https://doi.org/10.1158/0008-5472.CAN-08-4511
102. Lin CC, Tu CF, Yen MC, Chen MC, Hsieh WJ, Chang WC et al (2007) Inhibitor of heat-shock protein 90 enhances the antitumor effect of DNA vaccine targeting clients of heat-shock protein. Mol Ther 15:404–410
103. Lampen MH, van Hall T (2011) Strategies to counteract MHC-I defects in tumors. Curr Opin Immunol 23:293–298. https://doi.org/10.1016/j.coi.2010.12.005
104. Leone P, Shin EC, Perosa F, Vacca A, Dammacco F, Racanelli V (2013) MHC class I antigen processing and presenting machinery: organization, function, and defects in tumor cells. J Natl Cancer Inst 105:1172–1187. https://doi.org/10.1093/jnci/djt184
105. Seliger B (2014) The link between MHC class I abnormalities of tumors, oncogenes, tumor suppressor genes, and transcription factors. J Immunotoxicol 11:308–310. https://doi.org/10.3109/1547691X.2013.875084
106. Castilleja A, Ward NE, O'Brian CA, Swearingen B II, Swan E, Gillogly MA et al (2001) Accelerated HER-2 degradation enhances ovarian tumor recognition by CTL. Implications for tumor immunogenicity. Mol Cell Biochem 217:21–33
107. Businaro R, Profumo E, Tagliani A, Buttari B, Leone S, D'Amati G et al (2009) Heat-shock protein 90: a novel autoantigen in human carotid atherosclerosis. Atherosclerosis 207:74–83. https://doi.org/10.1016/j.atherosclerosis.2009.04.026
108. Direskeneli H, Ekşioglu-Demiralp E, Yavuz S, Ergun T, Shinnick T, Lehner T et al (2000) T cell responses to 60/65 kDa heat shock protein derived peptides in Turkish patients with Behçet's disease. J Rheumatol 27:708–713

109. Li R, Qian J, Zhang W, Fu W, Du J, Jiang H et al (2014c) Human heat shock protein-specific cytotoxic T lymphocytes display potent anti-tumour immunity in multiple myeloma. Br J Haematol 166:690–701. https://doi.org/10.1111/bjh.12943
110. Miconnet I, Salcedo M, Chouaib S, Lemonnier FA, Abastado JP, Kosmatopoulos K (2007) Induction of multiple $CD8^+$ T cell responses against the inducible Hsp70 employing an Hsp70 oligoepitope peptide. Oncol Rep 17:679–685
111. Van Herwijnen MJ, Van Der Zee R, Van Eden W, Broere F (2013) Heat shock proteins can be targets of regulatory T cells for therapeutic intervention in rheumatoid arthritis. Int J Hyperthermia 29:448–454. https://doi.org/10.3109/02656736.2013.811546
112. Song D, Chaerkady R, Tan AC, García-García E, Nalli A, Suárez-Gauthier A et al (2008) Antitumor activity and molecular effects of the novel heat shock protein 90 inhibitor, IPI-504, in pancreatic cancer. Mol Cancer Ther 7:3275–3284. https://doi.org/10.1158/1535-7163.MCT-08-0508
113. Yamano T, Mizukami S, Murata S, Chiba T, Tanaka K, Udono H (2008) Hsp90-mediated assembly of the 26 S proteasome is involved in major histocompatibility complex class I antigen processing. J Biol Chem 283:28060–28065. https://doi.org/10.1074/jbc.M803077200
114. Rajagopal D, Bal V, Mayor S, George A, Rath S (2006) A role for the Hsp90 molecular chaperone family in antigen presentation to T lymphocytes via major histocompatibility complex class II molecules. Eur J Immunol 36:828–841
115. Zhou D, Blum JS (2004) Presentation of cytosolic antigens via MHC class II molecules. Immunol Res 30:279–290
116. Maj T, Wei S, Welling T, Zou W (2013) T cells and costimulation in cancer. Cancer J 19:473–482. https://doi.org/10.1097/PPO.0000000000000002
117. Baumeister SH, Freeman GJ, Dranoff G, Sharpe AH (2016) Coinhibitory pathways in immunotherapy for cancer. Annu Rev Immunol 34:539–573. https://doi.org/10.1146/annurev-immunol-032414-112049
118. Berges C, Bedke T, Stuehler C, Khanna N, Zehnter S et al (2015) Combined PI3K/Akt and Hsp90 targeting synergistically suppresses essential functions of alloreactive T cells and increases Tregs. J Leukoc Biol 98:1091–1105. https://doi.org/10.1189/jlb.5A0814-413R
119. Collins CB, Aherne CM, Yeckes A, Pound K, Eltzschig HK, Jedlicka P et al (2013) Inhibition of N-terminal ATPase on HSP90 attenuates colitis through enhanced Treg function. Mucosal Immunol 6:960–971. https://doi.org/10.1038/mi.2012.134
120. Collins CB, Strassheim D, Aherne CM, Yeckes AR, Jedlicka P, de Zoeten EF (2014) Targeted inhibition of heat shock protein 90 suppresses tumor necrosis factor-α and ameliorates murine intestinal inflammation. Inflamm Bowel Dis 20:685–694. https://doi.org/10.1097/01.MIB.0000442839.28664.75
121. de Zoeten EF, Wang L, Butler K, Beier UH, Akimova T, Sai H et al (2011) Histone deacetylase 6 and heat shock protein 90 control the functions of Foxp3(+) T-regulatory cells. Mol Cell Biol 31:2066–2078. https://doi.org/10.1128/MCB.05155-11
122. Stenderup K, Rosada C, Gavillet B, Vuagniaux G, Dam TN (2014) Debio 0932, a new oral Hsp90 inhibitor, alleviates psoriasis in a xenograft transplantation model. Acta Derm Venereol 94(6):672. https://doi.org/10.2340/00015555-1838
123. Bae J, Munshi A, Li C, Samur M, Prabhala R, Mitsiades C et al (2013) Heat shock protein 90 is critical for regulation of phenotype and functional activity of human T lymphocytes and NK cells. J Immunol 190:1360–1371. https://doi.org/10.4049/jimmunol.1200593
124. Huyan T, Li Q, Dong DD, Yang H, Zhang J, Huang QS et al (2016) Heat shock protein 90 inhibitors induce functional inhibition of human natural killer cells in a dose-dependent manner. Immunopharmacol Immunotoxicol 38:77–86. https://doi.org/10.3109/08923973.2015.1119159
125. Trojandt S, Reske-Kunz AB, Bros M (2014) Geldanamycin-mediated inhibition of heat shock protein 90 partially activates dendritic cells, but interferes with their full maturation, accompanied by impaired upregulation of RelB. J Exp Clin Cancer Res 33:16. https://doi.org/10.1186/1756-9966-33-16

126. Tukaj S, Tiburzy B, Manz R, de Castro Marques A, Orosz A, Ludwig RJ et al (2014a) Immunomodulatory effects of heat shock protein 90 inhibition on humoral immune responses. Exp Dermatol 23:585–590. https://doi.org/10.1111/exd.12476
127. Tukaj S, Zillikens D, Kasperkiewicz M (2014b) Inhibitory effects of heat shock protein 90 blockade on proinflammatory human Th1 and Th17 cell subpopulations. J Inflamm (Lond) 11:10. https://doi.org/10.1186/1476-9255-11-10
128. Giannini A, Bijlmakers MJ (2004) Regulation of the Src family kinase Lck by Hsp90 and ubiquitination. Mol Cell Biol 24:5667–5676
129. Grune T, Jung T, Merker K, Davies KJ (2004) Decreased proteolysis caused by protein aggregates, inclusion bodies, plaques, lipofuscin, ceroid, and 'aggresomes' during oxidative stress, aging, and disease. Int J Biochem Cell Biol 36:2519–2530
130. Hayashi K, Kamikawa Y (2011) HSP90 is crucial for regulation of LAT expression in activated T cells. Mol Immunol 48:941–946. https://doi.org/10.1016/j.molimm.2010.12.014
131. Tsuji T, Matsuzaki J, Caballero OL, Jungbluth AA, Ritter G, Odunsi K et al (2012) Heat shock protein 90-mediated peptide-selective presentation of cytosolic tumor antigen for direct recognition of tumors by CD4+ T cells. J Immunol 188(8):3851. https://doi.org/10.4049/jimmunol.1103269
132. Salimu J, Spary LK, Al-Taei S, Clayton A, Mason MD, Staffurth J et al (2015) Cross-presentation of the oncofetal tumor antigen 5T4 from irradiated prostate cancer cellsDOUBLEHYPHENa key role for heat-shock protein 70 and receptor CD91. Cancer Immunol Res 3:678–688. https://doi.org/10.1158/2326-6066.CIR-14-0079
133. Bae J, Mitsiades C, Tai YT, Bertheau R, Shammas M, Batchu RB, Li C, Catley L, Prabhala R, Anderson KC, Munshi NC (2007) Phenotypic and functional effects of heat shock protein 90 inhibition on dendritic cells. J Immunol 178:7730–7737. PMID: 17548610
134. Bao R, Lai CJ, Qu H, Wang D, Yin L, Zifcak B et al (2009) CUDC-305, a novel synthetic HSP90 inhibitor with unique pharmacologic properties for cancer therapy. Clin Cancer Res 15:4046–4057. https://doi.org/10.1158/1078-0432.CCR-09-0152
135. Graham B, Curry J, Smyth T, Fazal L, Feltell R, Harada I et al (2012) The heat shock protein 90 inhibitor, AT13387, displays a long duration of action in vitro and in vivo in non-small cell lung cancer. Cancer Sci 103:522–527. https://doi.org/10.1111/j.1349-7006.2011.02191.x
136. Holland JP, Caldas-Lopes E, Divilov V, Longo VA, Taldone T, Zatorska D et al (2010) Measuring the pharmacodynamic effects of a novel Hsp90 inhibitor on HER2/neu expression in mice using Zr-DFO-trastuzumab. PLoS One 5:e8859. https://doi.org/10.1371/journal.pone.0008859
137. Sydor JR, Normant E, Pien CS, Porter JR, Ge J, Grenier L et al (2006) Development of 17-allylamino-17-demethoxygeldanamycin hydroquinone hydrochloride (IPI-504), an anti-cancer agent directed against Hsp90. Proc Natl Acad Sci U S A 103:17408–17413
138. Fogliatto G, Gianellini L, Brasca MG, Casale E, Ballinari D, Ciomei M et al (2013) NMS-E973, a novel synthetic inhibitor of Hsp90 with activity against multiple models of drug resistant target agents, including intracranial metastases. Clin Cancer Res 19:3520–3532. https://doi.org/10.1158/1078-0432.CCR-12-3512
139. Liu C, Workman CJ, Vignali DA (2016) Targeting regulatory T cells in tumors. FEBS J 283:2731–2748. https://doi.org/10.1111/febs.13656
140. Mendler AN, Hu B, Prinz PU, Kreutz M, Gottfried E, Noessner E (2012) Tumor lactic acidosis suppresses CTL function by inhibition of p38 and JNK/c-Jun activation. Int J Cancer 131:633–640. https://doi.org/10.1002/ijc.26410
141. Pyzer AR, Cole L, Rosenblatt J, Avigan DE (2016) Myeloid-derived suppressor cells as effectors of immune suppression in cancer. Int J Cancer 139:1915–1926. https://doi.org/10.1002/ijc.30232
142. Sitkovsky MV, Kjaergaard J, Lukashev D, Ohta A (2008) Hypoxia-adenosinergic immunosuppression: tumor protection by T regulatory cells and cancerous tissue hypoxia. Clin Cancer Res 14:5947–5952. https://doi.org/10.1158/1078-0432.CCR-08-0229

143. Conejo-Garcia JR, Benencia F, Courreges MC, Kang E, Mohamed-Hadley A, Buckanovich RJ et al (2004) Tumor-infiltrating dendritic cell precursors recruited by a β-defensin contribute to vasculogenesis under the influence of VEGF-A. Nat Med 10:950–958
144. Goel S, Wong AH, Jain RK (2012) Vascular normalization as a therapeutic strategy for malignant and nonmalignant disease. Cold Spring Harb Perspect Med 2:a006486. https://doi.org/10.1101/cshperspect.a006486
145. Gottfried E, Kreutz M, Haffner S, Holler E, Iacobelli M, Andreesen R et al (2007) Differentiation of human tumour-associated dendritic cells into endothelial-like cells: an alternative pathway of tumour angiogenesis. Scand J Immunol 65:329–335
146. Jain RK (2005) Normalization of tumor vasculature: an emerging concept in antiangiogenic therapy. Science 307:58–62
147. Stockmann C, Schadendorf D, Klose R, Helfrich I (2014) The impact of the immune system on tumor: angiogenesis and vascular remodeling. Front Oncol 4:69. https://doi.org/10.3389/fonc.2014.00069
148. Yang L, DeBusk LM, Fukuda K, Fingleton B, Green-Jarvis B, Shyr Y et al (2004) Expansion of myeloid immune suppressor $Gr^{+}CD11b^{+}$ cells in tumor-bearing host directly promotes tumor angiogenesis. Cancer Cell 6:409–421
149. Huang Y, Goel S, Duda DG, Fukumura D, Jain RK (2013) Vascular normalization as an emerging strategy to enhance cancer immunotherapy. Cancer Res 73:2943–2948. https://doi.org/10.1158/0008-5472.CAN-12-4354
150. Topalian SL, Drake CG, Pardoll DM (2015) Immune checkpoint blockade: a common denominator approach to cancer therapy. Cancer Cell 27:450–461. https://doi.org/10.1016/j.ccell.2015.03.001
151. Bali P, Pranpat M, Bradner J, Balasis M, Fiskus W, Guo F et al (2005) Inhibition of histone deacetylase 6 acetylates and disrupts the chaperone function of heat shock protein 90: a novel basis for antileukemia activity of histone deacetylase inhibitors. J Biol Chem 280:26729–26734
152. Ganai SA (2016) Histone deacetylase inhibitors modulating non-epigenetic players: the novel molecular targets for therapeutic intervention. Curr Drug Targets. PubMed PMID: 27231104
153. Seidel C, Schnekenburger M, Mazumder A, Teiten MH, Kirsch G, Dicato M et al (2016) 4-Hydroxybenzoic acid derivatives as HDAC6-specific inhibitors modulating microtubular structure and HSP90α chaperone activity against prostate cancer. Biochem Pharmacol 99:31–52. https://doi.org/10.1016/j.bcp.2015.11.005
154. Kroesen M, Gielen P, Brok IC, Armandari I, Hoogerbrugge PM, Adema GJ (2014) HDAC inhibitors and immunotherapy; a double edged sword? Oncotarget 5:6558–6572
155. Schotterl S, Brennenstuhl H, Naumann U (2015) Modulation of immune responses by histone deacetylase inhibitors. Crit Rev Oncog 20:139–154
156. Shen L, Orillion A, Pili R (2016) Histone deacetylase inhibitors as immunomodulators in cancer therapeutics. Epigenomics 8:415–428. https://doi.org/10.2217/epi.15.118
157. Gameiro SR, Malamas AS, Tsang KY, Ferrone S, Hodge JW (2016) Inhibitors of histone deacetylase 1 reverse the immune evasion phenotype to enhance T-cell mediated lysis of prostate and breast carcinoma cells. Oncotarget 7:7390–7402. https://doi.org/10.18632/oncotarget.7180
158. Lisiero DN, Soto H, Everson RG, Liau LM, Prins RM (2014) The histone deacetylase inhibitor, LBH589, promotes the systemic cytokine and effector responses of adoptively transferred $CD8^{+}$ T cells. J Immunother Cancer 2:8. https://doi.org/10.1186/2051-1426-2-8
159. Tiper IV, Webb TJ (2016) Histone deacetylase inhibitors enhance CD1d-dependent NKT cell responses to lymphoma. Cancer Immunol Immunother 65:1411. PubMed PMID: 27614429
160. Cao K, Wang G, Li W, Zhang L, Wang R, Huang Y et al (2015) Histone deacetylase inhibitors prevent activation-induced cell death and promote anti-tumor immunity. Oncogene 34:5960–5970. https://doi.org/10.1038/onc.2015.46
161. Lee SY, Huang Z, Kang TH, Soong RS, Knoff J, Axenfeld E et al (2013) Histone deacetylase inhibitor AR-42 enhances E7-specific $CD8^{+}$ T cell-mediated antitumor immunity induced by therapeutic HPV DNA vaccination. J Mol Med (Berl) 91:1221–1231

162. Song DG, Ye Q, Santoro S, Fang C, Best A, Powell DJ Jr (2013) Chimeric NKG2D CAR-expressing T cell-mediated attack of human ovarian cancer is enhanced by histone deacetylase inhibition. Hum Gene Ther 24:295–305. https://doi.org/10.1089/hum.2012.143
163. Woods DM, Sodré AL, Villagra A, Sarnaik A, Sotomayor EM, Weber J (2015) HDAC inhibition upregulates PD-1 ligands in melanoma and augments immunotherapy with PD-1 blockade. Cancer Immunol Res 3:1375–1385. https://doi.org/10.1158/2326-6066.CIR-15-0077-T

# Index


**A**
Adaptive immune responses, 77
Adoptive cellular therapy, 165
Allogeneic gp96 vaccine, 84, 85
Allosteric inhibitors, 26–27, 29–31
Analogous inhibitors, 24
Androgen receptor (AR) transcriptional activity, 33
Antigen-presenting cell (APC), 77, 78, 142, 145, 146, 150
Anti-infectious disease approach, 84
Anti-infectious disease vaccine approach, 83
Antisense oligonucleotides, 33–34
Antitumor vaccination
  allogeneic gp96, 84, 85
  combined therapeutic approach, 86
Antitumor vaccine approach, 83
ATP-competitive inhibitors, 29

**B**
Bcl-2-associated athanogene (BAG) protein, 12
Benzoquinone ansamycin inhibitors, 22–24

**C**
C terminus of Hsc70-interacting protein (CHIP), 13
Cancer, 76, 77, 79, 86
Cartilage oligomeric matrix protein (COMP), 104
CD91, 47, 63–69, 77, 78, 87, 90, 125, 130, 142, 145, 146, 150, 160–163, 167, 169
CD8 cytotoxic T lymphocyte (CTL), 79
$CD8^+$ T cell (CTL), 124
Cellular communication
  chemical gradients, 44
  ECV, 45
  gap junctions, 44
  immune synapse, 44
  intercellular, 44
  ligand-receptor interaction, 44
  T-, B-, and antigen-presenting cells, 44
  type of, 44
Chaperone
  cellular stress, 63
  DnaK system, 8
  entropic pulling method, 9 (*see also* Peptide substrates model)
  ECM regulation, 102
  ER, 104–106
  function, HSPs, 63
  Hsp40, 108, 109
  Hsps, 100
  "kinetic partitioning" model, 7, 8
  peptides, 64
Chaperone inhibitors
  cancer and neurodegenerative maladies, 22
  Hsp70 and Hsp90, 21
  protein crystallography, 25
Clusterin, 110
Clusterin, stress-inducible small Hsp, 33–34
Co-chaperones
  ATP and ADP-to-ATP exchange, 11
  BAG proteins, 12
  CHIP, 13
  client transfer, ATPase regulation and conformational dynamics, 21
  EEVD motifs, 13
  glucocorticoid receptor, 13




R. J. Binder, P. K. Srivastava (eds.), *Heat Shock Proteins in the Immune System*,
https://doi.org/10.1007/978-3-319-69042-1

Co-chaperones (*cont.*)
GrpE homologs, 12
Hip, 13
Hsp110/Hsp170 proteins, 12
J-domain-containing Pam16-Pam18 complex, 12
JDPs, 11
in vitro and in vivo complexes, 12
Colon carcinogenesis, 133
Concanavalin A (Con A), 127
C-terminal Hsp-interacting protein (CHIP), 30
Cytotoxic T lymphocytes (CTL), 161, 165, 166

**D**
Dendritic cells, 64
Diethylnitrosamine (DENA) treatment, 132

**E**
Enhanced immunity, 159, 168, 169
Entropic pulling method, 9
Evolution, HSPs
class Ia and class Ib genes, 148
gp96 and hsp70, evolutionarily conservation, 145
Ib- binding patterns and conservation, 151
*X. laevis* tumor immunity model, 145
Extracellular HSP (eHSPs)
amyloid β aggregation, 48
CD91-dependent endocytosis, 64
central nervous system, 48
in clinical conditions, 47
controlled therapy-induced tumor cell death, 162
eHSP10, 162
function, 46
Grp75 (HSPA9), 46
Grp78 (HSPA5) and Grp94 (HSPC4), 46
Hsp70, 46, 47
Hsp90α (HSPC3), 46
Hsp90β (HSPC4), 46
immune modulation, 160–162
immune responses, 64, 65
mechanisms of export, 49, 50
multi-factorial role, 160, 161
NK cell activity, 68
in pathological conditions, 47, 48
potential receptors, 47
in rheumatoid arthritis, 66
tumor exosome (TEX)-associated eHSPs, 162
in tumor immunosurveillance, 66–68
Extracellular matrix (ECM), 100–102, 104
BM, 98
chaperones, 102
clusterin, 110
fibroblasts, 98
Hsp/molecular chaperones, 103
Hsp40, 106–109
Hsp70, 105, 106
Hsp90
cytosolic isoforms, 100
extracellular, 101, 102, 104
Grp94, 104
interstitial/stromal, 98
protein and function, 98
SPARC, 111
structural support and cell signaling, 98
tumour microenvironment, 99
Extracellular vesicles (ECV)
classification, 51
Hsp70, 52
with HSP export, 50
in stress conditions, 51
signaling proteins/microRNAs, 45
as zip codes, 45

**F**
Fibronectin (FN), 101, 102
Functional cycle, Hsp70 proteins, 5, 6

**G**
gp96
activity, 127–128
enhanced Treg function, 126
gp96-based therapeutic vaccine against HBV, 131
gp96-Ig vaccine approach, 84–89
antitumor vaccination
allogeneic tumor cells, 84, 85
combined therapeutic approach, 86
cancer, 79
ER, 81
KDEL sequence, 80, 81
malaria vaccines, 90
mice model, 81
peptide-binding, 81, 82
perforin, 79
SIV/HIV
antibody production, 88
CD91, 87
electroporation (EP), 89
293-gp96$^{SIV}$Ig, 87
mucosal SIVmac251 infection, 89
tumorigenicity, 80
gp96-mediated T Cell responses, 130
gp96-mediated Treg activation, 126–127

gp96$^{SIV}$Ig and gp120 protein, 89
Grp94
  allosteric inhibitors, 26–27
  endoplasmic reticulum (ER)-restricted Hsp90 paralog, 26

**H**
Haematopoietic stem cells (HSC), 104
Heat shock protein (HSPs), 81, 87, 99, 100
  active secretion, mechanisms of, 65
  ATP-induced substrate release, 7, 8 (*see also* Chaperones; Co-chaperones)
  as chaperones, 63, 64
  destabilization, HSP clients, 160
  discovery, 45
  DnaK interactions, 6
  eukaryotic organisms, 6
  evolutionarily conservation, 145
  functional cycle, 5–6
  functions of, 4
  hallmark pathways, 159
  hallmarks of cancer, 169
  HSP70 and gp96, 142
  Hsp70 family, 45
  HSP90 client protein HER2, 159
  molecular mechanisms, 5
  NBD subdomains, 7
  partially folded substrates, 4
  SBD, secondary structure, 6, 7
  SBD-substrate interactions, 6
  stress and normal conditions, 4
  stress conditions and macromolecular crowding, 4
  stress-inducible, 45
  tumor cell overexpression, 158
Hemagglutinin (HA), 128
Hepatitis B virus (HBV), 126
Hepatocellular carcinoma (HCC), 131–133
Herpes simplex virus (HSV)-2, 131
Histone deacetylase inhibitors (HDACi), 170, 171
Hsp40
  Hsp47 and ECM, 108, 109
  MRJ/DNAJB6, 107
Hsp47, 108, 109
Hsp70
  ER-resident, 106
  extracellular, 105
Hsp90, 101, 102, 104
  extracellular
    cell migration, 104
    FN, 101, 102
  Grp94, 104
HSP-adjuvanted BCG vaccines, 129–130
HSP-adjuvanted human papillomavirus (HPV) vaccines, 129
HSP-adjuvanted influenza vaccine, 128
HSP70-based therapeutic vaccine against HSV-2, 131
Hsp90 co-chaperone inhibitors, 27–28
Hsp90 C-terminal inhibitors, 25–26
Hsp40, DnaJ/J-domain proteins, 32
HSP gp96 as target in chronic HBV infection, 132–133
HSP gp96 as target in inflammation-associated colon cancer, 133
Hsp70/Hsc70 inhibitors
  allosteric inhibitors, 29–31
  ATP-competitive inhibitors, 29
  subdomains, 28
  types of, 28
HSP inhibitors (HSPi)
  anti-tumor immunity, enhancement, 168, 169
  in B, T and NK cells, dose-dependent manner, 166, 167
  geldanamycin (GA), 165
  HSP90i, 165, 166
  in vitro and in vivo, 165
  modulations, in tumor immunogenicity, 163
  pro-oncogenic signaling, 166
  on tumor cell expression, co-stimulatory, 166
Hsp70-interacting protein (Hip), 13
Hsp90 N-terminal inhibitors, 22
Hsp90 second-generation inhibitors, 24–25
Human immunodeficiency viruses (HIV), 89

**I**
Immune system
  *Xenopus laevis*, 142, 143
Immunogenic cell death (ICD), 160, 161
Immunosurveillance, 66, 68
Immunotherapy, 77, 85, 86, 168–171
Innate immunity, 77
Innate T (iT) cells
  class Ib-restricted effector cells, 149
  iNKT cells, 148
  in tumor immunity, 147
  *Xenopus* XNC10-restricted iT cells, 149, 150
Intercellular communication, 44

**K**
"Kinetic partitioning" model, 7, 8

**L**

Lipopolysaccharide (LPS), 132

**M**

Malaria vaccines, 90
Mammalian relative of DNAJ (MRJ), 107
microRNA (miR)-29, 109
Mortalin, 46
Muramyl dipeptide (MDP), 132

**N**

Natural killer T-cells (iNKT) cells, 148, 149
Neuraminidase (NA), 128
Nitric oxide synthase (NOS2), 160
N-terminal domain (NTD), 22, 23

**P**

Peptide aptamers, 32
Peptide substrates model
  cellular and developmental signals, 11
  clathrin-coated endocytic vesicles, 11
  dimerization, 10
  DnaK binding, 10
  fibril depolymerization, 10
  homodimers/heteromeric complex, 10
  Hsp70-substrate interactions, 9
  protein disaggregation, 9
  toroidal hexameric AAA+ proteins, 9
  UPR regulator, 10
  viral infections, 11
  virion assembly, role of, 11
Perforin, 79
pH1N1 vaccine, 128
Placenta-derived gp96, 134–135
Pregnancy-associated plasma protein A (PAPP-A), 135
Progression-free survival (PFS), 158
Prostaglandin-$E_2$ ($PGE_2$), 159
Protein degradation
  Hsp70 system, 12
Protein deposition disorders (PDD), 100
Protein folding
  protein aggregation, 12
Protein transduction domain (PTD), 130

**Q**

Quality control network, 14

**R**

Radicicol inhibitors, 24
Receptor associated protein (RAP), 130
Regulatory cells, 162, 169, 170

**S**

Signaling
  extracellular HSP, 45
  proteins/microRNAs, 45
Small heat shock proteins (sHsps), 110
Smooth muscle cells (SMCs), 105
SPARC/osteonectin, 111
Stress observation system, 51
Stress response
  and macromolecular crowding, 4
  proteins misfolding, 9
  UPR receptors, 10

**T**

T regs, 65
T-cell receptor (TCR), 149
Toll-like receptors (TLRs), 125
Trans-activator of transcription (TAT) protein, 130
Tuberculosis (TB), 129
Tumor immunity, 66, 68
  innate T cells, 147, 148
  NK cell-mediated cytotoxicity, 147
  *Xenopus* nonclassical MHC class Ib (XNC) genes, 148, 149
  *X. laevis* tumor immunity model, 145
Tumor immunosurveillance, 66–68
Tumor microenvironment (TME), 165–171
Tumor necrosis factor (TNF), 132

**U**

Unconventional T cells, 142, 144, 145, 149, 151
Unfolded protein response (UPR), 10
Urokinase plasminogen activator receptor (uPAR), 107

**V**

Vaccines, 162, 165, 168, 170, 171
  CD8+ cytotoxic T cells, 77
  DNA-based, 89
  immune responses, 77

immunity, 76
immunogenicity and tolerability, 76
infectious disease, 77
Vascular normalization (VN), 169, 170

**X**

*Xenopus laevis*
cell surface marker (CTX), 144
immune system, 142, 143
lymphoid tumors and tumor immunity, 143, 144
peritoneal macrophage (pMac), 146
tumor immunogenicity, hsp70 isoform, 145, 147
XNC10-restricted iT cells, 149

**Z**

Zymography, 103

Printed by Printforce, the Netherlands